Pulses and Vegetables

UNDERUTILIZED CROP SERIES

Series co-ordinator: N. Haq

Interests in the use of underutilzed crop species are governed by many different concerns in developed and underdeveloped economies, these interests forming a continuum related to crop diversification, production of value-added products, rural development and increased income for growers and processors.

For too long, the world has relied heavily on a relatively small number of crops grown across wide areas and there have been many justifiable environmental concerns expressed. There is an increasing international awareness in the importance of investigating options for the development of new and alternative crops, set against a backdrop of worrying statistics on the loss of species diversity, in particular the loss of species which have not been looked at with a view of commercial exploitation as new crops.

The Underutilzed Crop Series is a short series of books, each covering a range of crops within a particular group which have received both local and international attention over the past decade and where enhanced research attention has been justified. Each book in the series has been drawn together by the International Centre for Underutilzed Crops (ICUC), the series co-ordinator is Dr N. Haq of ICUC. The first two books in the series are edited by Professor J.T. Williams of the International Fund for Agricultural Research.

This series is an invaluable source of reference for agricultural and plant scientists, nutritionists, development agencies and all those involved in research and policy decisions concerning agricultural development, biodiversity exploitation and so on.

Titles in this series

Pulses and Vegetables
Edited by J. T. Williams

Cereals and Pseudocereals
Edited by J. T. Williams

Forthcoming titles

Fruit and Nuts

Industrial Crops

Pulses and Vegetables

Edited by

J.T. Williams

International Fund for Agricultural Research, Washington D.C. metro, USA

CHAPMAN & HALL

London · Glasgow · New York · Tokyo · Melbourne · Madras

Published by Chapman & Hall, 2–6 Boundary Row, London SE1 8HN

Chapman & Hall, 2–6 Boundary Row, London SE1 8HN, UK

Blackie Academic & Professional, Wester Cleddens Road, Bishopbriggs, Glasgow G64 2NZ, UK

Chapman & Hall Inc., 29 West 35th Street, New York NY10001, USA

Chapman & Hall Japan, Thomson Publishing Japan, Hirakawacho Nemoto Building, 6F, 1–7–11 Hirakawa-cho, Chiyoda-ku, Tokyo 102, Japan

Chapman & Hall Australia, Thomas Nelson Australia, 102 Dodds Street, South Melbourne, Victoria 3205, Australia

Chapman & Hall India, R. Seshadri, 32 Second Main Road, CIT East, Madras 600 035, India

First edition 1993

Typeset in 10/12pt Palatino by ROM-Data Corporation Limited, Falmouth, Cornwall, England
Printed in Great Britain by St Edmundsbury Press, Bury St Edmunds, Suffolk

ISBN 0 412 46610 4

A catalogue record for this book is available from the British Library

Library of Congress Cataloging-in-Publication data available

∞ Printed on permanent acid-free text paper, manufactured in accordance with the proposed ANSI/NISO Z 39.48–199X and ANSI Z 39.48–1984

Contents

Contributors

S. Azam-Ali
School of Agriculture,
University of Nottingham,
Sutton Bonington, Loughborough,
Leics.
LE12 5RD,
UK.

T. Badra
Box 232,
Ste. 202,
3148 Kingston Road,
Scarborough,
Ontario,
MIM 1P4,
Canada.

M.J. Balick
New York Botanical Gardens,
Bronx,
New York,
NY 10458,
USA.

M.L.A. Bovi
Instituo Agonômica,
Av. Barão de Itapura 1481,
Caixa Postal 28,
13001 Campinas,
BRAZIL.

T.E. Carter Jr.,
USDA/ARS,
North Carolina State University,
Box 7631,
Raleigh,
NC 27695–7631,
USA.

R.I. Denton
International Jute Organization,
95A Road,
No. 4 Banani,
PO Box 6073,
Gulshan,
Dhaka-1213,
Bangladesh.

M. Grum
Research Division,
Ministry of Agriculture and Forestry,
PO Box 14,
Nuku'alofa,
Tonga.

M.P. Guerra
Universidade Federal de Sta. Catarina,
Campus Universitario S/N,
Trinidad,
Florianopolis,
SC 88049,
Brazil.

N. Haq
Department of Biology,
School of Biological Sciences,
Southampton University,
Southampton,
SO9 3TU,
UK.

A.R. Linnemann
Dept. of Tropical Crop Science,
Wageningen Agricultural University,
PO Box 341,
6700 AH Wageningen,
Netherlands.

R.E. Paull
Dept. of Plant Molecular Physiology,
University of Hawaii at Manoa,
Honolulu,
HI 96822,
USA.

E. Pimienta-Barrios
Universidad de Guadelajara,
Facultad de Ciencias Biologicas,
Blvd. Marcelino Garciá Barragán y Corregidora 44410,
Guadelajara, Jal.,
Mexico.

S. Shanmugasundaram
Asian Vegetable Research and Development Center,
PO Box 42,
Shanhua,
Tainan 74199,
Taiwan.

M. Sørensen,
Dept. Botany, Dendrology and Forest Genetics,
Royal Veterinary and Agricultural University,
Rolghedsvej 23,
DK-1958 Frederiksberg, C,
Denmark.

P.C. Tabora Jr.
Honduran Foundation for Agricultural Research,
Apartado Postal 2067,
San Pedro Sula,
Honduras.

A. Venthou-Dumaine
INRA des Antilles et de la Guyane,
BP 1232,
94185 Pointe-à-Pitre Cédex,
Guadeloupe,
French West Indies.

V. Vaillant
Laboratoire de Physiologie et Biochemie Végétales,
INRA BP 1232,
94185 Pointe-à-Pitre Cédex,
Guadeloupe,
French West Indies.

J.T. Williams
International Fund for Agricultural Research,
1611 N. Kent Street,
Suite 600,
Arlington,
VA 22209,
USA.

C. Zinsou
INRA des Antilles et de la Guyane,
BP 1232,
94185 Pointe-à-Pitre Cédex,
Guadeloupe,
French West Indies.

Introduction: underutilized pulses and vegetables

J.T. Williams

Interests in the use of underutilized species are spurred by quite different reasons in developed and developing economies, unless they relate to a major international market niche, such as plant extractive products. The interests form a continuum related to crop diversification, production of value-added products, rural development and increased income for growers and processors.

Additionally, the types of technologies available differ widely from sophisticated production and processing – including biotechnology – to traditional technologies which may or may not be appropriate.

In the developed world, it is easy to understand why research and development on processing staple crops for new end-uses is supported, or why a limited number of new crops became favoured, such as kenaf for the paper industry or crops for specific vegetable oil composition. With overproduction of staple crops in developed countries, the uses change: cereals end up as animal products rather than being wholly utilized directly as food (Moncke, 1983). There is greater innovation in crops to satisfy new needs from agriculture which the consumers recognize and there are new expectations of the industrial and food industries (Thomas, 1989).

Whereas similar expectations are likely to develop over time in the newly industrialized countries, in the least developed countries (LDC) requirements differ. On the one hand, many national plans of the past couple of decades are outdated since they underestimated population growth and in many LDCs the carrying capacity of the land is approaching its limits with demand for food growing faster than production. By the year 2000, 15 cities in Asia alone will have populations over 10 million. On the other hand, there are major challenges for sustainable agricultural development in the

Pulses and Vegetables. Edited by J. T. Williams. Published by Chapman & Hall,
2–6 Boundary Row, London SE1 8HN. ISBN 0 412 46610 4.

context of less environmental degradation and a slowing of the depletion of natural resources. Major challenges face agricultural research systems in the LDCs especially when national and international financial resources allocated to agricultural research and development are declining (Bonte-Friedheim, 1991). This being the case, it is hardly surprising that past exercises proposing research for LDCs on many underutilized crops have rarely been put into place nor received the support that is, in some cases, justified. Yet there are cases for wider use of a number of minor crops; and a groundswell of interest over the past decade led to the founding of the International Centre for Underutilized Crops (ICUC) to provide leadership and co-ordination on very specific priorities.

Also, in most developing countries there are large differences in the income levels of the different sectors of the population. The concerns of donor agencies relate largely to areas which target that part of the population which lives in poverty. This frequently contrasts with development at the national level which is often closely linked to urbanization and its problems. Cock and Lynam (1991) have eloquently pointed out that a dilemma exists, on the one hand, between small-scale farmers producing low-value, own-price inelastic goods and, on the other hand, farmers selling new products that are more own-price elastic. In part, this dilemma can be resolved by the production of new crops or innovative processing of old products into new forms. All too often, scientific interests of developed-country scientists pay scant attention to such dilemmas and better partnerships with colleagues in LDCs would be extremely fruitful in the face of major social transformations of the largely agrarian societies in developing countries.

Careful planning for the enhanced production of a limited number of priority underutilized crops can have a major and powerful impact on rural development. As in the case of cassava, discussed by Cock and Lynam (1991), targets should include accelerated rural development and also contribute to increasing demands for urban goods and services. At the outset a number of concerns need to be addressed. Long experience of heavy concentration on plantation crops, mostly for export, often favours the urban sector rather than enhancing income generation in the rural sector. In relation to the development strategies, it would be most sensible to look to new agricultural products which provide enhanced income for the rural sector which in turn interacts with rural demand for low technology consumer goods.

ENVIRONMENTAL CONCERNS

Environmental concerns – especially since the 1987 report of the World Commission on Environment and Development – with a number of global reports such as *Caring for the Earth: A Strategy for Sustainable Living* (a successor to the 1980 *World Conservation Strategy*), and the *Global Biodiversity Strategy* (WRI, IUCN, UNEP 1992) have led to much wider considerations of applications of ecological concepts to agriculture, appreciation of the highly complex smallholder farming systems characterized by a fragile, heterogeneous agro-sociological environment (Kesseba, 1989) and the need for sustainable international development strategies as a response to environmental, economic and social concerns. Achieving sustainable agriculture in the face of human population growth will require enhancement of the world's productive capacity as well as conservation of the resource base (NRC 1991, Walsh 1991).

Many of the environmental concerns have led to the promulgation, often on slender evidence, of enhanced attention to a range of traditional crops, especially when this appears to be an investment in biological diversity, for which there are loud clamours for conservation. But many of the concerns relate to the fact that numerous traditional crops have, for generations, been produced with low inputs and often on soils and in areas which are somewhat marginal for more productive agriculture. Sustainable agriculture may have been what our ancestors practised but the small farmers of this century cannot be expected to do only what their ancestors did: they need technology.

These are, of course, generalizations and gross oversimplifications, although there is clear evidence for the need in many parts of the tropics for a role for some traditional, often underutilized, crops in local-agricultural systems or woody crops in these or new agroforestry systems. Nonetheless, the calls for action often relate to mistaken concepts that equate traditional with sustainable, rather than attaining balanced production at higher levels of productivity which must be the aim for expanded production of any underutilized species.

Alternative crops may be totally new in any area or be existing minor crops of the area which require, for one reason or another, more attention. Whether they are indeed underutilized or not depends on their local context (see for instance, Okigbo, 1986). There are hundreds of crop plant species recorded as underutilized and to which some scientist or group has accorded a potential. In selecting a limited number, for this book, of vegetables in the broadest sense, a hard look was taken at those which had received both local and international attention over the past decade and where enhanced research attention has been justified. They represent, with the exception of winged bean, the handful which have emerged from the 16 or more vegetable and legume species (food crops and pulses) which were widely recognized as underutilized (NAS 1975, 1979). Others, but by no means all, may well emerge in the immediate future as other research programmes receive support (Diversity, 1990).

Environmental concerns would add to these the bewildering array of the thousands of useful plants gathered from the wild or 'protected' at the village level (Prescott-Allen and Prescott-Allen, 1982; McNeely *et al.*, 1990). Additionally, there is a major linkage between forestry and the contributions of forest resources to nutrition (FAO, 1989). Nonetheless, due consideration of most of these species will have to await environmental concerns mobilizing funding for research rather than the agricultural sector.

ENHANCED RESEARCH ATTENTION

What determines focus of the scientific community on any particular underutilized crop? In all probability, focus represents a synthesis of four major factors. First of all, new crops which have 'made it', such as those species introduced to new parts of the world after the age of Columbus, were often already major domesticates in their ancestral homes and provided better products in their new ones – whether they were potato, maize, bean and tomato from the New World, or banana, wheat and cabbage from the Old World. Others such as vanilla, rubber or oil palm filled a niche for new industries or enhanced international markets, e.g. sugar cane and spices.

Second, the long history of domestication and local selection *in situ* had already produced landraces or selected forms which were able to adapt relatively quickly. This is a far cry from the selection of a more or less wild useful plant and not only domesticating it but bringing it into agriculture. The investment required for the latter is huge and it is also high-risk and must be demand-led (Smith, 1988).

Third, the new crop has to fit into cropping systems, unless it is to be used largely on marginal, non-agricultural land. This is more complex than it seems, including farmer acceptance, commercial attention and adequate service infrastructure.

Fourth, in many cases, the germplasm of a target species available and ready to hand, can be very limited; even totally unsuitable for further adaptation or enhancement and the sheer practicalities of collecting a wide enough array for experimentation can be beyond any scientist's resources. Whereas the genetic heterogeneity of limited introductions may have been adequate in some cases (such as rubber and oil palm in the past) much of the 'hit or miss' of plant introductions has related to the diversity of the individual accessions introduced.

These factors, when duly considered, lead to the need for strategic planning, something the individual involved scientist is often not experienced in or indeed able to participate in. Hence, the major hope for any research programme to be sustained is to build up capacity in this area and this involves the assessment of a number of policy options (Williams and Haq, 1992). The current extent of new crops research and development in countries such as the Netherlands and India reflect such clear strategic planning (Jolliff, 1990; Mal, 1988). In this way, scarce financial resources for research can be targeted to the specific constraint, whether to the poor farmer or to industry. There is little to be gained from targeting such resources to existing intellectual research, of the type all too frequent in universities, unless the strategy for worthwhile pay-offs have been worked out and the planning has taken due cognizance of farmers.

Resource allocation and mobilizing support rests squarely on the shoulders of leaders of the national agricultural research systems (ISNAR, 1988) and in a scenario which assumes no change in present policies, developing countries will continue to favour the production of food grains and export crops. In scenarios which revise present policies, the door opens for planning such as that cited by Mal (1988) for India.

NETWORKING INFORMATION

It is a sad fact that much of the relevant information on potential new crops or the further exploitation of local ones is lost in countless reports which are not readily available at the developing country level. These reports deal with trials of an adaptive nature, agronomy, germplasm sources, failures of plant introductions and similar matters. The limited number of economic plant databases do not provide the type of information required and there is a long way to go before it will be available (Williams, 1989). ICUC has started work in this area, but the best method will be to build up the networks of involved scientists who have common goals and shared interests, and who determine the priorities for action.

Also, for many minor crops much of the information has been of little interest to scientists in developed countries. Yet the most important daily protein source in many tropical countries, where 75% of the world's pulses are grown, comes from tropical legumes. The other main source of protein is local leafy vegetables, similarly of interest in the tropics, hence the emphasis in this book.

GERMPLASM COLLECTIONS

For the species dealt with through major accounts in this book, the germplasm collections have been built up because of research interest in the particular crop. Some remain as research collections, e.g. that for yam bean, but others such as bambara groundnut and lupins are genetic resources collections associated with specific collection, conservation and documentation.

Consideration of germplasm raises a number of issues which probably need pragmatic solutions on a crop-by-crop basis. Whereas a major international effort spearheaded by the Consultative Group on International Agricultural Research (CGIAR) has allocated resources for genetic conservation of major staple crops, the comprehensiveness of minor crop germplasm collections is very patchy. Some of the hitherto underutilized crops which are well along the road to wider exploitation continue to have inadequate germplasm collections; others such as okra and *Amaranthus* are better served. The range of differences in the comprehensiveness of these collections reflects the degree of importance attached to them by researchers and users and priorities accorded in the late 1970s by the International Board for Plant Genetic Resources (Grubben, 1977) and specific programmes for tropical vegetables being initiated for a limited number accompanied by status reports and action programmes, e.g. on amaranths, okra, West African yams. Even so, more attention was paid to vegetable crops such as tomato, *Allium*, *Capsicum* and sweet potato.

Many minor crops are, of course, in the same situation, with regard to incomplete attention to genetic resources, and since the early 1980s, others have been added to the lists for action with little actual practical result: minor species and both crops and woody species are now considered important in agroforestry situations, but again little targeted collection or documentation nor indeed strategy has been developed (Williams, 1991) despite the importance of these in developing countries.

Integrated genetic resources programmes have focused heavily on more major crops. Although a recent analysis (Prescott-Allen and Prescott-Allen, 1990) showed that data available relate to 103 crops species which in fact feed the world (the bulk of calories still come from around 25), there are still many crops of local importance for which production figures do not exist. Most of the underutilized priority crops fall into this latter category. Global strategy for the *ca.* 100 crops listed (Prescott-Allen and Prescott-Allen, 1990) leaves many for which no comprehensive genetic conservation mechanism has been planned. How then will underutilized species be dealt with?

For the crops described in this book, there are few, if any, international breeding efforts. Whereas some of the research, as evidenced by the accounts, is indeed international or supported by donor funding, much of the adaptive research is the responsibility of national scientists and ultimately the genetic conservation systems will be the responsibilities of national programmes. Many of these are in a relatively weak stage of development and it is unlikely that we shall see fully comprehensive and co-operative germplasm systems in place for these crops for some time.

The key factor will be the user-driven requirements for the availability of germplasm and the maintenance of sufficient diversity to sustain enhancement. The crops are not being used in the same way that major staples are being used to provide the bulk of the world's food. Apart from the need for conservation where there is clear evidence of genetic erosion, many underutilized crops are likely to remain in the agroecosystems where they have proved useful. Advanced germplasm systems are only needed when germplasm enhancement is warranted and active as in the case of the examples in this book.

Since enhanced use of underutilized species is often closely linked to rural development, new partnerships with responsible non-governmental organizations with multiple constituencies could be a useful adjunct to national programmes.

THE FUTURE

The species described in this book are largely those of value in the tropics and subtropics and the arid zones. From a nutritional point of view, they cover crops which are increasingly important as a complement to carbohydrate staples. They are crops with existing niches in agroecological systems and which, with some creativity – and viability – in national research organizations, can take a quantum jump in productivity and quality from the traditional agriculture systems of which they have long been a part.

Due in a large part to the creation of CGIAR and the network of internationally-funded agricultural research centres working on food research problems of the developing world, research towards future food production during the next doubling of the world's population is in place. For staple foods, traditional systems of agriculture will become more and more transformed to modern productive systems. Although great progress has been made in the past 20 years towards developing a global research effort to attack agricultural problems of the tropics and subtropics, much remains to be done at the national level. Some countries such as those with large research systems might well continue some efforts on underutilized crops but the small developing countries with few scientists will still need strong partnerships with international efforts or other countries (Oram, 1990).

Reorientation of international research towards sustainability will address issues such as protection of the genetic base for agriculture, preserving the natural resource base, researching the problems pertinent to less favourable environments and external inputs (CGIAR, 1990). In the face of the need to produce food for expanding populations, little such reorientation will be towards 'new' crops unless they fit specific multiple cropping needs, strategies to arrest degradation of fragile lands so as to foster sustained production of food, fuel and income (NRC, 1991), strategies to use saline water or soils (NRC, 1990) or other very specific needs. Some attention to new or underutilized crops will, of course, continue in the context of current farming systems research or soil management collaborative research (for an example in the humid tropics, see Colfer, 1991). And increasingly diversity will **not** always improve sustainability; there is the need for carefully designed agroecosystems. Particularly at the village level, the underutilized species will continue to provide complementary items of nutrition, particularly protective nutrients, and will continue to be important components of home gardens.

REFERENCES

Bonte-Friedheim, C. (1991) *NARS in a New Age*, 6pp. International Service for National Agricultural Research, The Hague.

CGIAR (1990) Sustainable agricultural production. Final report of the CGIAR Committee, Doc. MT/90/18, Consultative Group on International Agricultural Research, World Bank, Washington, DC.

Cock, J.H. and Lynam, J. (1991) Research for development. *Developing World Agriculture*, Grosvenor Press International, pp. 87–95.

Colfer, C.J. (1991) *Towards Sustainable Agriculture in the Humid Tropics and Building on the TropSoils Experience in Indonesia*. TropSoils Bulletin no. 91–92, North Carolina State University.

Diversity (1990) New plant diversity research effort launched at Nigerian symposium. Diversity, **6** (3/4) 7–8.

FAO (1989) *Forestry and Nutrition: A Reference Manual*. Food and Agriculture Organization of the United Nations, Rome, Italy.

Grubben, G.J.H. (1977) *Tropical Vegetables and their Genetic Resources*, International Board for Plant Genetic Resources, Rome, Italy.

ISNAR (1988) *The Changing Dynamics of Global Agriculture*, Proceedings of a Seminar on Research Policy Implications for National Agricultural Research Systems, International Service for National Agricultural Research, The Hague.

Jolliff, G.D. (1990) Strategic planning for new-crop development, in *Advances in New Crops*, (eds J. Janick and J.E. Simon), Timber Press, Portland, Oregon, pp. 29–40.

Kesseba, A.M. (ed.) (1989) *Technology Systems for Small Farmers*, Westview Press, Boulder, Colorado.

Mal, B. (1988) Underutilized plants programme in India: Concept and future perspective, in *Life Support Plant Species*, (eds R.S. Paroda, P. Kapoor, R.K. Arora and B. Mall), National Bureau of Plant Genetic Resources, New Delhi, pp. 153–9.

McNeely, J.A., Miller, K.R., Reid, W.V., *et al.* (1990) *Conserving the World's Biological Diversity*, IUCN/WRI/CI/WWF-US/World Bank, Gland, Switzerland.

Moncke, E.A. (1983) *International Grain Trade 1950–1980*, Agriculture Experiment Station, Tech. Bulletin 247, University of Arizona, Tucson.

NAS (1975) *Underexploited Tropical Plants with Promising Economic Value*, National Academy of Sciences, Washington, DC.

NAS (1979) *Tropical Legumes: Resources for the Future*, National Academy of Sciences, Washington, DC.

NRC (1990) *Saline Agriculture: Salt-Tolerant Plants for Developing Countries*, National Academy Press, Washington, DC.

NRC (1991) *Towards Sustainability, A Plan for Collaborative Research in Agriculture and Natural Resource Management*, National Academy Press, Washington, DC.

Okigbo, B.N. (1986) Broadening the food base in Africa: The potential of traditional food plants. *Food and Nutrition*, **12**, 4–17.

Oram, P. (1990) Agricultural productivity, growth and the structure and organization of agricultural research, in *Technology Policy for Sustainable Agricultural Growth*, IFPRI Policy Briefs 7, International Food Policy Research Institute, Washington, DC, pp. 7–9.

Prescott-Allen, R. and Prescott-Allen C. (1982) *What's Wildlife Worth? Economic Contributions of Wild Plants and Animals to Developing Countries*, International Institute for Environment and Development, Earthscan, London.

Prescott-Allen, R. and Prescott-Allen, C. (1990) How many plants feed the world? *Conservation Biology*, **4**, 365–76.

Smith, R.W. (1988) The place of life support species in hostile or risk-prone environments: an overview, in *Life Support Species: Diversity and Conservation*, (eds R.S. Paroda, P. Kapoor, R.K. Arora and B. Mal), National Bureau of Plant Genetic Resources, New Delhi, pp. 14–19.

Thomas, T. (1989) Food industry and agriculture, In *New Crops for Food and Industry*, (eds G.E. Wickens, N. Haq and P. Day), Chapman and Hall, London, pp. 13–22.

Walsh, J. (1991) *Preserving the Options: Food Production and Sustainability*, CGIAR Issues in Agriculture No. 2, Consultative Group on International Agricultural Research, Washington, DC.

Williams, J.T. (1989) Plant introduction and international responsibilities, in *New Crops for Food and Industry*, (eds G.E. Wickens, N. Haq and P. Day), Chapman and Hall, London, pp. 345–51.

Williams, J.T. (1991) Plant genetic resources: some new directions. *Adv. Agron.*, **45**, 61–91.

Williams, J.T. and Haq, N. (199?) Some policy aspects of new crops, in *Proc. Symposium on New Crops*, October, 1991 in Indianapolis (in press).

WRI, IUCN, UNEP (1992) *Global Biodiversity Strategy*, World Resources Institute, Washington, DC.

Part One

Crops Benefiting from Major Research

1

Bambara groundnut (*Vigna subterranea*)

Anita R. Linnemann and Sayed Azam-Ali

1.1 HISTORY AND USES

The bambara groundnut (*Vigna subterranea* (L.) Verdc.) is an indigenous African crop grown primarily for its seeds which are eaten fresh when semi-ripe, as a pulse when dry and mature, or ground into flour. The common English name appears to be derived from a tribe of agriculturalists, the Bambara, who nowadays live mainly in Mali.

The centre of origin of bambara groundnut has been a matter of dispute for many years. Guillemin, Perrottet and Richard reported the probable occurrence of wild plants in Senegal in 1832. However, this observation was never confirmed, and Hepper (1970) suggested that plants on abandoned farmland in that area had been mistaken for wild ones. In 1909 Dalziel found the plant supposedly wild near Yola in northern Nigeria, and, in the same year, Ledermann made a similar discovery near Garua in Cameroon (Dalziel, 1937; Stapf, 1912). Harms (1912), however, had considered these examples to be subspontaneous and not wild ancestors of the cultivated forms. The issue was resolved in 1957 by Hepper (1963), who collected small seeded forms of bambara groundnut from uncultivated savanna land between Yola and Garua. Plants grown from the collected seeds had a slender, trailing habit, which contrasts sharply with the compact and sturdy habit of cultivated types. Hepper concluded that the area of distribution of wild progenitors is from the Jos Plateau and Yola in northern Nigeria eastwards to Garua in Cameroon, and possibly as far as the Central African Republic.

For many centuries bambara groundnut has been cultivated in the tropical regions of Africa south of the Sahara where it is indigenous. Major producers are Nigeria, Niger, Ghana, Haute Volta and Côte d'Ivoire, but it is widely grown in eastern Africa and Madagascar. Slaves took the crop to

Pulses and Vegetables. Edited by J. T. Williams. Published by Chapman & Hall, 2–6 Boundary Row, London SE1 8HN. ISBN 0 412 46610 4.

Surinam; in 1762 seeds were collected in the vicinity of Paramaribo and sent to Linnaeus in Uppsala for description (Ostendorf, 1962). Marcgrav de Liebstad (1648) reported the crop from Brazil in the 17th century and it is now found in parts of South and Central America. The crop was taken to Asia, particularly India, Indonesia, Malaysia, Philippines and Sri Lanka. It is found in parts of Northern Australia.

1.1.1 **Uses**

Bambara groundnut is predominantly grown for human consumption. Immature seeds are consumed fresh or grilled. Seeds can be boiled in the immature, green state, either shelled or unshelled, until quite soft. The kernels are eaten either by themselves or mixed with immature groundnut or maize (Johnson, 1968).

Ripe, shelled nuts are very hard and cannot be eaten without cooking, often for longer than other species (Ezueh, 1977). They are also difficult to grind (Smartt, 1959) and are either pounded to flour and then boiled to a stiff porridge or soaked or split and then boiled. The porridge remains palatable for a long time and is often used as a food for long journeys (Holm and Marloth, 1940). It is also preserved as a winter relish.

An alternative is to roast, boil, crush and eat the seeds as a relish with maize meal porridge ('sadza'). Often, the dried seeds are made into a paste which forms the basis of various fried or steamed products such as 'akara' and 'moin-moin' (Obizoba, 1983).

In Nigeria, the bambara groundnut paste is also often used as a steamed product ('okpa') after being wrapped in fresh banana or plantain leaves and boiled for 30–50 min (Obizoba, 1983).

The boiled and salted whole kernels are frequently offered as appetizers in restaurants in Angola and Mozambique (Johnson, 1968). Hepper (1970) reported that roasted bambara groundnut meal has occasionally been used as a coffee substitute.

Bambara groundnut seeds can also be used as a feed for pigs and poultry (Oluyemi *et al.*, 1976) and the haulm as fodder (Doku and Karikari, 1971a). Oluyemi *et al.* (1976) successfully included raw and autoclaved bambara groundnut seed at 40% of the basal ration for chicks. The performance of the animals was significantly improved with the autoclaved seed compared with the raw material, largely because of the denaturation of protease inhibitors and the increased availability of metabolizable energy. Oluyemi *et al.* (1976) also reported that the efficiency of utilization of gross energy in bambara groundnut was superior to that of autoclaved cowpea seeds and acacia nuts.

Owusu-Domfeh *et al.* (1970) reported that the true biological value and protein efficiency ratio of autoclaved and raw bambara groundnut were significantly higher than those of groundnut (*Arachis hypogaea* L.) cake and cowpea (*Vigna unguiculata* (L.) Walp). However, Uwaegbute (1977) demonstrated a loss of feeding value with cooked bambara groundnut seeds when fed to chicks. This effect was rectified by supplementing raw and cooked beans with 0.2% lysine and methionine, which resulted in a significant increase in growth and feed conversion efficiency.

1.2 TAXONOMY AND ECOLOGY

Bambara groundnut belongs to the Leguminosae, subfamily Papilionoideae. Linnaeus described the plant in *Species Plantarum* (1763) and named it *Glycine subterranea*. In the 19th century Du Petit-Thouars (1806) changed the generic name to *Voandzeia*. This was derived from the Malagasy name 'voanjo', which was composed of 'voa' (seed) and 'anjo' (that satisfies well) (Rassel, 1960). In 1980 the name was changed from *V. subterranea* (L.) Thouars to *Vigna subterranea* (L.) Verdc. after detailed taxonomic studies (Maréchal *et al.*, 1978; Verdcourt, 1980).

Hepper (1970) distinguished two botanical varieties: var. *spontanea* (Harms) Hepper, which includes the slender and diffuse wild forms of bambara groundnut, and var. *subterranea*, for the more robust cultivated forms.

There are no cultivars of bambara groundnut in the strict sense, but locally farmers, who cultivate it in small plots, distinguish different seed lots by their appearance. In the Chivi communal area in Zimbabwe, for instance, farmers discern at least eight different types to which they have given names such as 'Svutu' (meaning 'suit'). The heterogeneous primitive cultivars tend to be trailing, better selected ones are more clumped; however, whatever selection has taken place has been casual and over long periods of traditional cultivation. Due to their very localized nature they can hardly be considered landraces and little is known about wide adaptability.

Bambara groundnut is cultivated at elevations up to 1600 m. It is typically a short-day plant. For successful cultivation a frost-free period of at least 3.5 months is necessary. The plant grows better in bright sunshine and under average day temperatures of 20–28°C. Successful cultivation also requires an evenly distributed, moderate rainfall from sowing time until flowering. Although bambara groundnut can be cultivated in savanna areas with a seasonal rainfall of *ca.* 600–750 mm, for optimum yields an annual rainfall of 750–900 mm is necessary. The crop tolerates heavy rainfall, except at maturity. Additionally there are forms adapted to the cooler, moist highlands of Zimbabwe, and to areas too arid for other crops. Bambara groundnut will grow on any well-drained soil, but light sandy loams of pH 5.0–6.5 are most suitable. The crop yields moderately well on poor sandy soils marginal for groundnut and other pulses, and it is one of the best crops for savanna ochrosols (Brammer, cited by Doku and Karikari, 1971a). Soils rich in nitrogen tend to encourage vegetative growth at the expense of pod production (Johnson, 1968) and are to be avoided. Calcareous soils are reported to be unsuitable, but those rich in phosphorus and potassium would be beneficial (Hepper, 1970). In Indonesia it is also cultivated on lowland peats (Driessen and Permadhy Sudewo, 1977).

Figure 1.1 Bambara groundnut (*Vigna subterranea* (L.) Verd.). 1, Habit of flowering plant; 2, flower; 3, fruits; 4, seed. From L.J.G. van der Maesen and S. Somaatmadja (eds) (1989) *Plant Resources of South-East Asia No. 1. Pulses*, Pudoc, Wageningen, p. 75, reproduced with permission.

1.3 MORPHOLOGY AND ANATOMY

Bambara groundnut is an indeterminate, annual herb up to 30 cm in height with creeping, much-branched, leafy lateral stems just above ground level (Figure 1.1).

The plant has a well developed tap root with lateral roots on the lower part. In association with *Rhizobium*, the roots form rounded and sometimes lobed nodules.

Lateral branches may develop from buds in the axils of the cotyledons, which remain underground (hypogeal germination). Differences in the length of internodes result in bunched, intermediate and spreading types. Cultivated forms usually have a bunched or intermediate growth habit. Some forms produce adventitious roots at the nodes of the stems. Leaves are pinnately trifoliate, glabrous, with erect, grooved petioles thickened at the base. Each leaf has four stipels with extrafloral nectaries; two stipels subtend the terminal leaflet, while the lateral leaflets have one each. The narrowly elliptic to lanceolate leaflets are attached to the rachis with marked pulvini. The petiolules of the first two to three leaves are very short; but from the third to fourth leaf the petiolule of the terminal leaflet is longer than those of the lateral leaflets. The terminal leaflet, often larger than the lateral leaflets, may reach a length of 10 cm and a width of 3.5 cm.

The racemes consist of 1–3 (usually 2) whitish to yellow flowers on a short axillary peduncle. The stamens are diadelphous, nine with filaments fused for more than half their length and one free vexillary stamen. The style is bent and bears a tuft of hairs near the terminal stigma. Self-pollination is normal. After pollination and fertilization the peduncle, which has a rounded, glandular apex, lengthens, thus bringing the ovaries to or just below ground level.

The pods usually develop underground, and may reach up to about 2.5 cm in diameter. When mature, the pericarp of the indehiscent pods is hard and wrinkled. Pods are usually almost spherical and contain one round to oblong seed, although they may occasionally contain up to four seeds (Rassel, 1960). Seeds are very variable in colour: white, cream, yellow, red, purple, brown or black; the coloration is uniform, mottled, blotched or striped (Figure 1.2); some have an eye with a dark colour around the white hilum. Seed colour, size and hardness characterize farmers' local cultivars along with the plant form: spreading, bunched or semi-bunched. The anatomical features of the leaves have been studied in relation to water availability. Nishitani et al. (1984) showed that the adaxial epidermis had nearly 100 stomata per mm^2 and these stomata were always located above a respiration cavity. There were about 200 stomata per mm^2 in the abaxial epidermis. Almost all adaxial epidermal cells and most abaxial cells were rectangular water storage cells. Those in the adaxial epidermis were larger than those in the abaxial epidermis. The water storage cells accounted for about 35% of the leaf tissue.

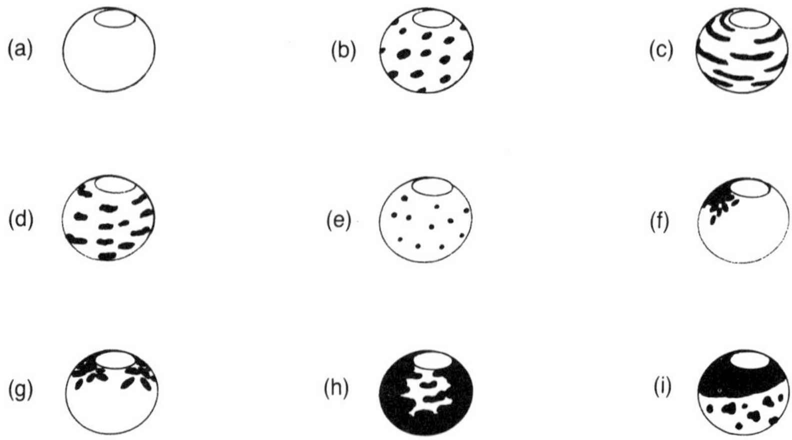

Figure 1.2 Testa patterns of bambara groundnut seeds (a) no pattern; (b) mottled; (c) striped; (d) marbled; (e) dotted; little rhomboid spotting on (f) one side of hilum; (g) both sides of hilum; (h) much rhomboid spotting on both sides of hilum; (i) Holstein pattern. Adapted from IBPGR/IITA/GTZ (1987).

Brown at the University of Nottingham recently measured leaf characters under different water regimes using controlled-environment glasshouses.

Table 1.1 shows that irrigation had a marked influence on leaf width,

Table 1.1 Dimensions of major constituent cells in transverse sections of bambara groundnut leaves at 150 days after sowing (DAS)

Irrigation treatment	*Wet*	*Medium*	*Dry*
Leaf width (μm)	293	236	213
stomatal number (no. mm)			
adaxial	64	118	141
abaxial	141	227	196
Guard cell dimensions (μm)			
adaxial	50 × 19	38 × 10	45 × 13
abaxial	25 × 16	38 × 13	18 × 13
Epidermal cell dimensions (μm)			
adaxial	55 × 35	45 × 30	33 × 23
abaxial	38 × 27	39 × 25	21 × 18
Palisade cell dimensions (μm)	48 × 12	45 × 13	28 × 13
spongy mesophyll dimensions (μm)	19 × 19	18 × 18	18 × 15

The treatments were: Wet, soil irrigated every week to field capacity; Medium, 40 mm irrigation applied at 62 DAS and 90 DAS; Dry, no irrigation after crop establishment (35 DAS).

stomatal distribution and number and on the size of the epidermal and palisade cells.

A remarkable feature of leaves in all treatments was the distribution of large bundle sheath cells around smaller evenly spaced vascular bundles. Although bambara groundnut is a C3 species, its leaf structure resembles that of classical C4 'Krantz' anatomy and this peculiar characteristic deserves further detailed study.

The testa of the seeds consists of an outer epidermis covered by a cuticle, a hypodermal layer of hourglass-shaped pillar cells with a distinct girdle of thickening in the central region of the radial walls (Rowson, 1952) and thin-walled parenchyma with some vascular tissue. Palisade cells with bottle-shaped lumina form the epidermis. In coloured seeds, pigment is found in the cavities of the epidermal cells. The parenchyma cells of the cotyledons contain oil drops, minute aleurone grains, and starch grains, and pronounced hila (Vaughan, 1970).

1.4 PHYSIOLOGY OF GROWTH AND DEVELOPMENT

1.4.1 **Germination**

The hypogeal germination of cultivated forms usually takes 7 to 15 days. The germination of the wild forms of var. *spontanea* is erratic and slow; from 26–31 days to indefinite.

Amuti and Pollard (1977b) found that from the start of germination onwards the amounts of stachyose and raffinose decreased and were virtually nil on the 12th day of germination. The amount of verbascose remained fairly constant. The sucrose content of germinating cotyledons increased, depending on genotype, to about 2.5 to 4 times the original value on the 7th day of germination. It remained high for about 10 days and then downturned again. Free glucose and fructose were found after the 7th day of germination. Free galactose (or galactose-containing oligosaccharides other than the ones initially present) was not detected at all in germinating cotyledons, although the raffinose oligosaccharides contain a galactose moiety. Bambara groundnut cotyledons convert galactose mostly to sucrose.

Umezurike and Numfor (1979) found that the starch content of cotyledons of bambara groundnut seeds decreased during the first 10 days of germination from about 51.8% to 6.8%. During the first four days the activity of the enzymes α-amylase and total amylase increased and thereafter declined. The activity of starch phosphorylase was low during the first six days of germination when the amylase activity was relatively high. A subsequent decrease in the amylase activity was accompanied by a substantial increase of the activity of starch phosphorylase, indicating an increasing importance of phosphorolysis in starch degradation at the later stages of germination. Starch phosphorylase could provide, as demonstrated in other crops, glucose-1-phosphate to form UDP-glucose for the synthesis of sucrose for the embryo. After germination the activities of enzymes belonging to the phosphogluconate pathway (hexokinase, glucose-6-phosphate dehydrogenase and 6-phosphogluconate dehydrogenase) increased steadily. The operation of this pathway is important since it may supply intermediates for biosynthetic processes.

During prolonged storage at ambient temperatures the composition of bambara groundnut seeds changes. This change is also associated with progressive loss of viability (Sreeramulu, 1983). Analysis of changes in food reserves during seed storage revealed only small changes up to 12 months preceding reduced germinability. The major decrease in seed viability at 12–18 months is associated with larger changes. The chemical components which showed a decrease during storage were starch, total soluble sugars, fat and insoluble-N; and increases were observed in the amount of soluble-N, amino-N and fatty acids. However, loss of viability is more likely to be the result of changes in endogenous growth regulators, in particular auxin, phenol, phenolic acids, growth inhibitors such as abscissic acid, and coumarin.

Under conditions of reduced moisture content and storage at low temperatures the seeds can be stored for reasonably long periods without major changes. Seeds for genetic conservation are kept under such conditions in genebanks.

1.4.2 **Flowering and pollination**

Flowering starts 30 to 55 days after sowing and may continue until the plant dies. Japanese researchers (Nishitani *et al.*, 1988a) found that flower clusters of genotype 'Bo' from Indonesia were visible 12 days before anthesis. The flowers reached their full size on the day before anthesis. Experiments were conducted to investigate the possibility of growing bambara groundnut under field conditions in the Kyushu region of Japan (Nishitani and Inouye, 1981) and hence information on day-length response was needed. In trials with two day lengths and three temperatures genotypes 'Bo' from Indonesia and 'D-25', 'H-13', 'H-48' and 'H-60' from the University of Ghana, Legon, produced flowers under short day (8 h light) and continuous light conditions. The number of days from sowing to flowering was very similar under both photoperiods. The genotypes from Ghana initiated flowers at constant temperatures of 20, 25 and 30°C, but 'Bo' only at 20 and 25°C. In further research (Nishitani *et al.*, 1988b) on an additional sixteen accessions from the germplasm collection of the International Institute of Tropical Agriculture (IITA) in Nigeria, two accessions, namely TVsu 310 and 634, failed to produce flowers under conditions of continuous light. Continuous light delayed flowering by 6 to 11.5 days in four accessions (TVsu 270, 312, 625 and 657). Flowering in bambara groundnut can apparently be day neutral, or show a quantitative or qualitative short-day response.

Flowers are complete. Pistil and stamina lie enclosed in the keel petals which never open completely. Pollen maturity and stigma receptivity both occur just before or soon after the standard and wing petals open. At this time pistil and stamina are almost of equal length, thus facilitating self-pollination (Doku and Karikari, 1971b). There has been some confusion as to whether some forms are self-pollinated and others cross-pollinated. In Ghana the more spreading genotypes yielded better in the presence of ants. At first this effect was attributed to cross-pollination brought about between different flowers of the same plant (geitonogamy) which would provide a stimulus for fruit development (Doku, 1969). Follow-up experiments (Doku and Karikari, 1971b) demonstrated that the increased yields of semi-bunched and spreading genotypes in the presence of ants were the result of the soil being loosened by the insects which made it easier for fertilized flowers to penetrate the soil. The attraction of ants to plants may be associated with the presence of extrafloral nectaries on the leaves. Presumably selection under cultivation gradually led to the development of more bunched genotypes in which the dependence on agents for fruit burial disappeared.

1.4.3 **Fruit set**

Fertilized flowers develop into fruits on or beneath the soil surface. Apparently, reproductive development is not inhibited by light, unlike groundnut where pegs must penetrate the soil surface before further development takes place. The pod grows first. It reaches its mature size about 30 days after fertilization. The seeds develop in the subsequent 10 days. They are mature when the parenchymatous lining of the shell has disappeared and brown patches appear on the outside of the pod (Doku and Karikari, 1970). Early maturing genotypes reach this stage *ca.* 90 days after sowing. Late maturing genotypes may require 150 days or more; mean temperature during the season influences the time taken to achieve physiological maturity. There is a tendency for bunch types to mature earlier than the spreading ones. Fruit development may be influenced by photoperiod. Trials of accessions from different locations in Africa, exposed to several day length treatments (12 h or less and 14 h or more) showed that long photoperiods delayed or even prevented fruit set in accessions with day neutral or delayed flowering, indicating a quantitative and a qualitative short-day response, respectively (Linnemann, 1991a). Specific day length requirements for successive stages of development have been found in other crops too. For photosensitive cowpea cultivars, for example, Steele *et al.* (1985) observed a range of short-day requirements for inflorescence initiation.

Some cultivars even needed shorter days before initiated inflorescences could expand and flower. Okra (*Abelmoschus* sp.) is an example of a non-leguminous plant which requires specific short photoperiods for successive development stages. Cultivar 'Perkins Long Pod' Ivorien, for example, has a critical photoperiod of between 13.30 and 14 h for flower initiation and of between 13 and 13.15 h for anthesis and fruit set (Siemonsma, 1982).

When staining techniques indicated no influence of photoperiod on pollen fertility of bambara groundnut, the ovaries of plants from short and long photoperiods were subjected to anatomical studies (Linnemann, 1991b). Embryos were found in ovaries from both long and short photoperiods. The embryos developed at the same rate up to 18 days after flowering; by then the average length of an embryo was *ca.* 0.03 mm. At this stage growth stopped under long photoperiods. The ovaries remained on the plant for another two to three weeks, although more and more of them aborted over time. In a trial in which plants were transferred from long photoperiods to inductive conditions of 12 h or less, some of the ovaries produced under long photoperiods developed mature seeds under the short photoperiods (Linnemann, 1991a). Thus photoregulation of development is an important trait in bambara groundnut, providing the plant with a flexible mechanism to adapt to circumstances that create seasonal fluctuations in the length of the growing period. Preliminary observations on West African accessions indicate that in general photosensitivity decreases with distance from the equator.

Bambara groundnut may produce one-seeded and double-seeded pods on the same plant. To clarify the reason for this, anatomical and embryological studies were carried out on genotype 'Bo' from Indonesia (Nishitani *et al.*, 1988a). At anthesis the average size of the ovary was 1.07 × 0.49 mm. Each ovary contained two ovules which remained similar in appearance during the development of the embryo sac. Double fertilization occurred on the day after anthesis. Three days after anthesis the first cell divisions of zygote and endosperm nuclei were observed, resulting in identical ball-type embryos at five days after anthesis. At seven days after anthesis the development of the ovules started to differ. That closest to the style continued to grow and matured at 70 days after anthesis. By then the ovule closest to the pedunculus was still a small embryo. The authors (Nishitani *et al.*, 1988a) give no explanation for the observed differences in development.

An assessment of endogenous growth regulators at 10, 20, 30, 40 and 50 days of seed development showed that the amounts of auxins and gibberellins were small in young seeds, increased up to 30 days of development and then rapidly declined until maturity (Sreeramulu, 1982). Throughout seed development auxin content was greater than the gibberellin content. The acidic inhibitor content was low during the first 20 days, then showed a fast increase up to 40 days, followed by a dramatic decline to nearly zero. The peak of acidic inhibitor content coincided with the peak in dry matter accumulation, which suggests that the inhibitor prevents precocious germination but not an increase in the amount of food reserves in the seed lobes.

1.4.4 **Influence of soil moisture stress**

Since the crop is adapted to a wide range of soils, thriving on lighter, well-drained ones, an understanding of soil moisture stress is important in upgrading the agronomy. Ameyaw and Doku (1983) performed trials at 30, 40, 50 and 75% available moisture (AM), where field capacity was considered 100% AM, achieved by watering to field capacity at intervals of 6, 5, 4, and 2 days, respectively. Increasing yields (measured in terms of seeds per plant and seed weights) were obtained in the following order: 75, 30, 50 and 40% AM; all differences were significant.

Elia and Mwandemele (1986) determined the experimental effects of water deficits. The highest number of pods per plant was obtained after daily watering. Decreased water availability reduced the number of flowers per plant, the percentage of fertile pollen, plant dry weight, plant height and number of branches per plant. These results show that, although bambara groundnut grows well in areas of marginal rainfall, yields are higher if there is sufficient water to improve seed setting.

To assess the potential of bambara groundnut for dryland agriculture in south-east Kenya, Hornetz (1988) investigated crop water requirements and applied this information to plan land use patterns using the WATBAL simulation model (Kutsch and Schuh, 1983). The accession used for these experiments turned out to require at least 75 days of sufficient water supply in order to give minimum yields in two out of three years. It was concluded that the cultivation of bambara groundnut would be possible during the second rainy season in parts of the Machakos district and in Voi, as well as during the first rains in Mariakani.

In 1988 and 1989 experiments in controlled-environment glasshouses (Nyamudeza, 1989; Lewitt, 1990) were carried out to further understand plant growth and moisture availability. Treatments ranged from an irrigated control, in which water was applied each week to return the soil to field capacity, to a stored water regime which received no irrigation from crop establishment until final harvest. Total evaporation, E_t, was estimated each week using a neutron probe; soil surface evaporation, E_s, was measured gravimetrically using trays of soil placed between the crop rows. Weekly and seasonal crop transpiration, E_p was calculated from (E_t - E_s). Total biomass accumulated was measured throughout the season using conventional growth analysis. The relation between accumulated biomass and transpiration was found to be linear for all treatments, with a slope of 4.0 g of dry matter per kg of transpired water. This 'dry matter/water use ratio' for total biomass was found to be identical to that of a groundnut variety in the same range of experimental conditions. However, although both species produced similar amounts of total biomass per unit of transpiration, there were clear differences in the proportion of biomass allocated to reproductive yield. For example, in the most drought-stressed treatment, groundnut failed to produce any pods whereas bambara groundnut produced a seed yield of 300 kg ha^{-1}.

It is still unclear how bambara groundnut is able to withstand greater water stress than other pulses and still produce stable yields. However, the relative sizes of above- and below-ground crop components may be influential. For example, in controlled-growth experiments the final number and individual expansion rates of leaves in bambara groundnut plants were less than in groundnut (Babiker, 1989; Karim, 1990), resulting in a smaller leaf area index and consequently a reduced demand for transpiration. Bambara groundnut also allocated a greater fraction of its total dry weight to roots than did groundnut in both irrigated and drought treatments (Nyamudeza, 1989). Thus, in relative terms, bambara groundnut appears to maintain both a greater supply of water to roots and a smaller demand for water by shoots. This strategy has clear advantages when water is scarce but may be a disadvantage when water is plentiful, because assimilates are allocated to roots rather than to photosynthetic surfaces or reproductive structures.

1.5 BIOCHEMICAL COMPOSITION

The main use of bambara groundnut is the seeds and these are often referred to as a complete food because they contain protein, carbohydrate and fat in sufficient proportions to provide a nutritious food (Poulter and Caygill, 1980). This more recent biochemical evidence confirms the validity of the original Malagasy name for the crop ('voanjo' – seeds that satisfy well) (Rassel, 1960).

Umezurike and Numfor (1979) found about 4.2% (w/w) soluble carbohydrates and 51.8% (w/w) starch in the cotyledons of bambara groundnut seeds before germination. The soluble reserve carbohydrates in seeds from Ghana consisted of the disaccharide sucrose (22.0%) and raffinose oligosaccharides: stachyose (44.6%), raffinose (31.1%) and a small amount of verbascose (2.3%) (Amuti and Pollard, 1977b). A genotype from Kenya also contained a trace of the raffinose oligosaccharide ajugose (Amuti and Pollard, 1977a). Free monosaccharides, like glucose, fructose and galactose, were not found in any of the ungerminated seeds these researchers studied.

Table 1.2 summarizes some reported ranges in the biochemical composition of bambara groundnut seeds. The seeds are not a significant source of lipids when compared with groundnut. Nevertheless, their lipid fraction compares favourably with some other pulses, such as pigeon pea (*Cajanus*

Table 1.2 Biochemical composition of bambara groundnut

	Gross energy	*Protein*	*Fat*	*Carbo-hydrate*	*Dietary fibre*	*Ash*
	(kJ)	(%)	(%)	(%)	(%)	(%)
Brough and Azam-Ali (1992)	1826	25.2	7.9	42.8	12.8	
FAO (1982)	1548	16.0	6.0	65.0		3.0
Linnemann (1987)	1540	18.0	6.0	60.0	3.3	
Oliveira (1976)	1732	19.2	7.8	69.3		3.6
Owusu-Domfeh *et al.* (1970)	1911	21.3	7.5	61.2		3.3
Oluyemi *et al.* (1976)	1869	22.8	6.1	61.8	4.4	4.8

cajan (L.) Millsp.) (1.2%), lentil (*Lens culinaris* Med.) (0.6%) and cowpea (1.6%) as reported by FAO (1982). The gross energy value of bambara groundnut seeds is greater than that of these other three pulses. Although bambara groundnut contains less oil than groundnut and soya bean (*Glycine max* (L.) Merr.), it has a higher carbohydrate value and its protein content is comparable with that of groundnut and many other legume species.

1.5.1 **Carbohydrate**

The carbohydrate fraction of bambara groundnut seeds is predominantly composed of starch and non-starch polysaccharides with lesser amounts of reducing and non-reducing sugars (Phillips and Abbey, 1989).

1.5.2 **Protein**

The large variation in reported protein levels (Table 1.2) indicates both genetic variation within the species and the interaction of genotypes with their environments. For example, the application of fertilizers, inoculation with *Rhizobium*, and increasing plant density can all increase protein yields but do not influence protein quality. On the other hand, applications of phosphorous and S-triazine herbicides increase protein and free amino acids, though their exact effects still have to be evaluated (Deshpande and Damodaran, 1990). The low protein yields of many subsistence pulses, such as bambara groundnut, is often attributed to genotypes having been selected for their ability to produce some yield under very poor conditions. Attention to improvement in crop management, varietal selection and plant breeding can increase both protein content and composition and probably more rapidly than similar research on more favoured pulses.

In general, the amino acid composition of bambara groundnut seeds is similar to that of many other pulses (Nwokolo, 1987) with a good balance

Table 1.3 Reported amino acid concentrations in bambara groundnut expressed as g/16 g N_2

	Brough (1991)	*Oliveira (1976)*	*Linnemann (1987)*	*FAO (1982)*	*Mosse & Pernollet (1982)*
Isoleucine	3.9	4.2	3.5–4.0	4.4	4.4
Leucine	7.6	6.9	7.4–7.7	7.8	7.8
Lysine	6.8	6.3	6.8–7.7	6.4	6.4
Methionine	1.4	1.4	1.2–1.3	1.8	1.8
Cysteine	1.3	1.1	1.2–1.4	1.0	1.1
Phenylalanine	5.6	6.4	5.4–5.8	5.6	5.6
Tyrosine	3.5		3.2–3.4	3.5	3.5
Threonine	3.5	3.0	3.2–3.6	3.5	3.5
Valine	4.8	6.1	4.4–4.8	5.3	5.3
Arginine	7.0	5.2	6.5–7.7		6.3
Histidine	2.9	2.4	3.1–3.5		3.0
Alanine	4.0		3.6–4.2		4.5
Asparagine	11.1		10.9–11.9		11.8
Glutamine	16.2		16.7–17.0		16.8
Glycine	3.6		3.6–3.8		4.0
Proline	4.5		4.1		5.1
Serine	5.2		5.1–5.4		5.6
Tryptophan		0.9		1.1	

of the essential amino acids (Salunkhe and Kadam, 1989). The range in amino acid concentrations reported by a number of workers is summarized in Table 1.3. Uwaegbute (1977) reported that bambara groundnut contains much more lysine than other legumes. However, the sulphur-containing amino acids methionine and cysteine are limiting (Owusu-Domfeh *et al.*, 1970; Linnemann, 1987). The quality of a protein can be chemically graded by comparing its amino acid content with that of a reference protein. Hen's egg protein is often recommended for this purpose. The resulting 'chemical score' provides an index of protein quality and a basis for inter-species comparison. The chemical score for bambara groundnut protein, calculated from FAO (1982) data, was 49, which compares favourably with that of other legumes, such as cowpea (40), tepary bean (*Phaseolus acutifolius* A. Gray) (43) and pigeon pea (21) and is similar to that of groundnut (52) and soya bean (50).

Nwokolo (1987) reported that amino acid availability in bambara groundnut (83%) is significantly higher than in that of African yam bean (*Sphenostylis stenocarpa* (A. Rich.) Harms)(63%). He considered that this information was relevant to nutrition field workers in Africa who use yam bean to formulate diets intended to meet the amino acid requirements of malnourished children and adults.

Table 1.4 Reported values of fatty acid content of bambara groundnut (expressed as %)

	Brough (1991)	*Linnemann* (1987)	*Gaydou* et al (1983)	*Kay* (1979)
Palmitic (16:0)	23.2	18–24	24.1	19.4
Palmitoleic (16:1)	1.0			
Stearic (18:0)	5.0	5–9	5.2	11.8
Oleic (18:1)	22.6	18–24	21.6	24.4
Linoleic (18:2)	39.0	35–40	40.4	34.2
Linolenic (18:3)	3.1	2–3	2.2	
Arachidic (20:0)	1.6	1–3	1.2	5.3
Eicosenoic (20:1)	0.8			
Behenic (22.0)	4.4	3–7	2.5	4.9
Lignoceric (24:0)		1–2	0.5	
Lauric (12:0)			2.4	1.0
Myristic (14:0)				2.0
Unsaponifiable matter of lipid fraction		1.6–2.4	2.4	

1.5.3 **Fatty acids**

Table 1.4 shows the range in composition of fatty acids in bambara groundnut seeds. The relative proportions of fatty acids are comparable with those of soya bean (Poulter, 1981). The essential fatty acids (linoleic and linolenic acid) account for between 37 and 43% of total fatty acid content. The approximate ratio of saturated to unsaturated fatty acids in bambara groundnut is 1:2.

1.5.4 **Anti-nutritional factors**

Most reports indicate the presence of a trypsin inhibitor in the seeds (Owusu-Domfeh *et al.*, 1970; Epias *et al.*, 1979; Griffiths, 1979; Poulter, 1981; Poulter and Caygill, 1980; Martino-Ferrer and Ferrer, 1983; Salunkhe and Kadam, 1989). Poulter (1981) reported that inhibitory activity ranged from 6.75 to 15.44 TIU mg^{-1} in 100 g samples of whole seed selected from eight genotypes.

Owusu-Domfeh *et al.* (1970) and Salunkhe and Kadam (1989) reported that the trypsin inhibitor is inactivated by autoclaving. Epias *et al.* (1979) and Griffiths (1979) indicated that although a substantial proportion of the trypsin inhibitor activity remained after heat treatment, total activity was reduced, suggesting the presence of a heat-stable (tannin) and heat-labile

Table 1.5 Reported mineral and vitamin composition of bambara groundnut expressed in mg/100 g dry matter

	Oliveira (1976)	*Linnemann* (1987)	*FAO* (1982)	*Ndiokwere* (1982)	*Nnadi* et al., (1976)	*Salunkhe and Kadam* (1989)
Minerals						
Phosphorus	345			384	550	380
Calcium	66	65	85	93	70	73
Potassium	1935			368	1570	
Magnesium	350				170	
Sodium	12					
Iron	8	6	4	48		8
Copper	1					
Zinc	8					
Manganese	15					
Vitamins						
Riboflavin	0.17	0.10				0.12
Niacin (actual)	2.14	2.00				2.1
Niacin (potential)	4.89					
Thiamine		0.30	0.18			0.28
Ascorbic acid						1.0
Retinol (IU/100 g)						30

(protein) factor. Poulter (1981) found that the tannin content of eight accessions ranged from 0.36 to 0.94 g/100 g. Further, he reported that trypsin inhibitory activity was directly correlated with the quantity of protein nitrogen extracted from each of the accessions, which, in turn, was found to be inversely related to their tannin contents. This activity was reduced by 85% following simple processing of the flour to form curds.

Tannins are located mainly in the seed coat and their concentration is correlated with seed colour as it is in common beans (*Phaseolus vulgaris* L.). Similarly, Poulter (1981) found the highest level of tannin was found in a bambara groundnut accession with brown/red seeds and the lowest tannin levels in accessions with cream-coloured seeds.

Martino-Ferrer and Ferrer (1983) isolated the trypsin inhibitor from ungerminated seeds and found that it existed in two forms, monomeric and dimeric, with molecular weights of 13 200 and 26 3000 in the denaturated and native forms respectively. Further studies on the kinetics of enzyme inhibition revealed that the enzyme – inhibitor complex is formed from two molecules of enzyme and one of inhibitor, with an intrinsic dissociation constant in the order of 1.28×10^{-8} M. No phytohaemagglutinins (Toms and Western, 1971), cyanogenic glucosides or alkaloids (Linnemann, 1987) have been found in bambara groundnut.

1.5.5 **Mineral and vitamin composition**

Table 1.5 summarizes the reported ranges in minerals and vitamins found in bambara groundnut seeds. Typical values for calcium and phosphorus are low compared with species such as cowpea and groundnut. With regard to the B vitamins, bambara groundnut contains less thiamine but comparatively more riboflavin than both cowpea and groundnut (Owusu-Domfeh *et al.*, 1970).

1.6 AGRONOMY, YIELD AND PRODUCTION

1.6.1 **Cropping systems**

As well as being grown in small plots, in many traditional farming systems bambara groundnut is grown as an intercrop in association with one or more other species (Figure 1.3).

Typically, these include cereals (millet, sorghum, or maize), root or tuber crops (cassava) other legumes (groundnut or cowpea) or vegetables (okra or pumpkin) (Doku, 1967; Doku and Karikari, 1971a, Ezueh, 1977; Haque, 1980; Linnemann, 1990). The crop may also be grown as a sole crop (Stanton *et al.*, 1966; Figure 1.4).

Figure 1.3 Bambara groundnut intercropped with cassava in Western Province, Zambia.

Figure 1.4 A woman's field of bambara groundnut near Yola, Nigeria.

In Nigeria the crop may be grown in rotation with maize, cowpea, cassava or yam (Okigbo, 1973). In Malawi there may be a 5 to 10 year rotation based on maize - groundnut - maize - common bean. After the first maize and groundnut crop, the following maize and common bean crop may be interplanted with cowpea or bambara groundnut (Wrigley, 1969).

Valentine (1963) reported that bambara groundnut has been successfully grown after tobacco, however this practice is not recommended where infestations of eelworm are likely.

Bambara groundnut is useful in crop rotations because it may improve the nitrogen status of the soil. In Zimbabwe, Mukurumbira (1985) reported that the crop had a higher residual nitrogen effect than groundnut, maize or fallow. Consequently, there was no need to apply nitrogen to maize when this was grown after bambara groundnut in the rotation. In contrast, nitrogen was still required when maize was grown after a previous crop of maize or after fallow.

Table 1.6 summarizes the residual effects of groundnut, bambara groundnut, fallow and a previous maize crop on the subsequent grain and yield of maize growing without applied nitrogen. Plots previously used for bambara groundnut produced significantly more grain than all the other treatments, although differences in stover yield were not significant. When the yield on the fallow plots was subtracted from the corresponding yields

Table 1.6 Residual effects of groundnut, bambara groundnut, maize and fallow on the grain and stover yields (kg ha^{-1}) of a subsequent maize crop

	Groundnut	*Bambara groundnut*	*Maize*	*Fallow*
Grain	6190	7560	3880	4280
Stover	9740	12050	6690	7840

Adapted from Mukurumbira, 1985.

on the groundnut, bambara groundnut and maize plots, the residual benefit of each crop could be quantified. Again bambara groundnut had a higher residual nitrogen effect on grain and stover production.

The residual contribution of bambara groundnut to a succeeding crop has also been investigated by Nnadi and Balasubramanian (1978), Nnadi (1980) and Nnadi *et al.* (1981). Total root nitrogen mineralized after 84 days of incubation of bambara groundnut roots in fallow soil was 48.7%. Liming reduced mineralized total root nitrogen to 12.4%. In terms of cultivation, the traditional method of hoe-ridging provided a greater fraction of mineralized nitrogen early in the season compared with zero tillage, bullock-ridging and tractor cultivation (Nnadi, 1980).

1.6.2 **Soil preparation**

Bambara groundnut prefers deep free-draining soils with a light, friable seedbed which is conducive to plants burying the pods (Johnson, 1968). It is recommended that the soil should be deep-ploughed to about 20 cm and brought to a fine tilth (Valentine, 1963). If ridging equipment is available, Johnson (1968) recommends making flat-topped ridges 90 cm apart if the soil is susceptible to waterlogging or is shallow. Otherwise, a flat seedbed is preferred.

1.6.3 **Sowing**

Since shelled seeds are reported to deteriorate at ambient temperatures (Brook *et al.*, 1977), they should be shelled as near to planting as possible in order to maximize viability. Depending on ambient temperatures and on relative humidity (RH), seed viability may persist for several months or over a year (Sreeramulu, 1983; Mabika, personal communication).

Although large seeds (more than 9.5 mm diameter) have been recommended for planting as they reportedly produce more vigorous seedlings (Karikari, 1969), in terms of fractional emergence, Zulu (1989) observed that across a wide range of soil moisture levels, small (less than 8.0 mm diameter) and medium (8.0–9.5 mm diameter) sized seeds consistently had a greater final emergence than large ones. He found that under irrigated conditions, small seeds of a Zimbabwean cultivar reached a maximum emergence of almost 90% whereas large seeds of the same cultivar reached a maximum of only about 65%. In addition to having a greater final emergence, small seeds also emerged at a faster rate. For example, in the same experiment small seeds reached 65% emergence after 10 days whereas the large seeds required 15 days to reach the same value.

The normal requirement for germination in natural conditions is between 7 and 15 days. Zulu (1989) found that maximum fractional germination, defined as an emerged radicle of 20 mm long, occurred at between 22 and 36°C in contrasting genotypes from Zimbabwe and Zambia. He found that no germination was evident at temperatures below about 12°C and above about 45°C. At 14°C germination started 96 h after sowing whereas at 33°C it started about 60 h earlier. However, within this range the rate of germination did not respond markedly to temperature. This is in contrast with the observed behaviour of many species which exhibit clearly defined minimum, optimum and maximum temperature for germination.

The poor response of germination rate to temperatures reported by Zulu (1989) may be a genetic trait, but it is more probable that an inherently linear response of germination rate to temperature is masked by the inability of seeds to imbibe sufficient water because of their hard seed coats. Further, although germination may already have occurred, the root radicle could have been prevented from emerging through the hard seed coat in time to be counted as 'germinated'. Zulu noted that the extending root radicle could often be seen coiling beneath the intact seed coat. Clearly, any means of reducing the influence of the seed coat on imbibition, germination and emergence is likely to positively effect the uniformity of crop establishment.

In order to investigate the germination rate of bambara groundnut under drought conditions, Zulu (1989) measured germination across a range of water potentials created by using various concentrations of polyethylene glycol. He found that germination was severely diminished even by small reductions in water potential. Thus, where the uptake of water is already restricted by the hard seed coat, any minor reductions in the external water potential can significantly restrict the overall uptake of water by the seed. It appears that, although bambara groundnut is a drought-resistant crop, an adequate supply of water at the germination stage is critical to successful crop establishment.

It is clear that for rapid and uniform crop establishment planting should ideally be completed early in the rainy season (Linnemann, 1987). This is usually between March and April in the western highlands of Cameroon, in July to August in the northern savanna zone of Cameroon (Westphal, 1981) and in November to December in Malawi (Nyasaland Protectorate, 1961) and Swaziland (Brook *et al.*, 1977). However, in practice, bambara groundnut is usually grown as a secondary food crop and is often planted very late after staple food crops, such as maize, and cash crops have been sown (Linnemann, 1990). Thus, planting dates vary considerably in any one place depending on the sequence of major crops. For example, in the Western Province of Zambia planting takes place between early November and early February (Linnemann, 1990). The periods during which bambara groundnut are sown are often, in practice, associated with relatively low soil moisture contents in the soil surface layers and this may be a major contributing factor for variability in seedling establishment often recorded.

Johnson (1968) reported that the timing of planting is critical to final yield. For Zimbabwe he recommended mid-November as the optimum sowing time and noted that delaying the sowing date by only three weeks from 30 November caused a yield reduction from 2530 kg ha^{-1} to 840 kg ha^{-1} in a clay soil and from 1420 kg ha^{-1} to 200 kg ha^{-1} in a sandy soil. Similarly, in Malawi seed yields were reduced from 2290 kg ha^{-1} for a crop sown on 14 December to 40 kg ha^{-1} for a crop sown on 8 February (Nyasaland Protectorate, 1961).

Phased planting is also practised. For example, this is prevalent in Sukumaland, Tanzania, where rainfall is irregular in amount and distribution (Ruthenberg, 1980).

Before sowing, it is recommended that seeds are treated with Thiram (tetramethylthiuram disulfide) at the rate of 1 kg per 120 kg of seed (Johnson, 1968).

1.6.4 **Sowing rates and depths**

The 1000-seed weight for bambara groundnut is between 500 and 750 g (Kay, 1979). FAO (1961) reports the average sowing rate is 35 kg ha^{-1} for Tanzania, 25–45 kg ha^{-1} for Kenya, 50–60 kg ha^{-1} for South Africa and 60–75 kg ha^{-1} for eastern Nigeria.

Rassel (1960) reported a typical planting density of 6.3 plants m^{-2} arranged in a 40 × 40 cm spacing. Johnson (1968) reported that dense populations are essential for good yields and recommended planting seeds either 10 or 15 cm apart in 45 cm rows; these would provide planting densities of about 22 and 14 plants m^{-2} respectively. Recent research indicates that even denser populations may be required to maximize the amount of radiation intercepted by the crop. For example, a plant population of 29 plants m^{-2} achieved a maximum of 90% fractional light interception only after about 100 days of growth (Karim, 1990). Thus, a significant proportion of total solar radiation must have passed through the crop canopy without being usefully intercepted by leaves for photosynthesis. However, it should be noted that these high planting densities are only appropriate in conditions where water stress or poor soil fertility do not limit crop productivity.

When planted as a sole crop one seed is usually sown per station. Rassel (1960) reported that in intercropping systems up to four seeds may be sown per station. However, this practice is probably rare, as farmers tend to make economical use of material for planting.

Bambara groundnut is usually planted by hand, with or without a hoe, and in these circumstances the depth of sowing may be quite variable. Johnson (1968) reported that a maize planter with thick plates and enlarged holes is also suitable for mechanical planting.

Optimum sowing depth depends on soil type. On heavy soil seeds are sown about 2.5 cm deep while on light soils a sowing depth of 5–7.5 cm is recommended (Doku and Karikari, 1971a). Under wet conditions shallow sowing has been reported to achieve better and earlier emergence (Stanton *et al.*, 1966).

Zulu (1989) investigated the effect of sowing depth on the emergence of seedlings. He found that sowing depths as great as 12 cm had no effect on final emergence. In contrast, under the same experimental conditions the emergence of groundnut seedlings was impaired at sowing depths greater than 8 cm. The relative insensitivity to sowing depth of bambara groundnut compared to groundnut may be associated with differences in their modes of emergence. Groundnut is epigeal and the hypocotyl must therefore pull two bulky cotyledons through the soil. In contrast, the sole organ that has to emerge in bambara groundnut is the epicotyl.

1.6.5 *Rhizobium*

Symbiotic nitrogen fixation takes place with bacteria of the cowpea inoculation group (Doku, 1969; Somasegaran *et al.*, 1990; Thompson and Dennis, 1977) with which bambara groundnut nodulates freely. For example, in a study of 23 strains of *Rhizobium*, Somasegaran *et al.* (1990) reported that *Bradyrhizobium* strain TAL 169, isolated from cowpea, formed the most effective symbiotic relationship with bambara groundnut when tested under glasshouse conditions in Hawaii. Moreover, TAL 169 was consistently effective on all 10 genotypes tested in that study.

In Senegal, Gueye and Bordeleau (1988) tested six *Rhizobium* strains on 24 genotypes and found that both indigenous and introduced (from NifTAL) *Rhizobium* strains nodulated all entries. In their study, the indigenous *Rhizobium* strain MAO 113 was found to form the most effective association with all genotypes, outperforming even TAL 169.

Other *Rhizobium* strains that are effective in promoting growth have also been found in experiments at NifTAL (National Research Council, 1979).

Brooks *et al.* (1988) investigated the effectiveness of six elite NifTAL and 13 indigenous West African strains of *Rhizobium* on accessions of bambara groundnut from Ghana and Togo. They found that five of the six NifTAL strains and two of the West African strains had effective nodulating potential when compared with uninoculated controls. In some cases the fixation rate even produced greater yield increases than were achieved by applying nitrogen fertilizer.

A nitrogen fixation rate of 106 kg ha^{-1} was reported for bambara groundnut by Dennis (1977).

Inoculation with soil from an old field in which the crop has been grown is recommended when planting a new field (Stanton *et al.*, 1966). Good soil drainage and sufficient amounts of phosphorus are essential for both the main and lateral roots to produce the typically small, rounded or lobed nodules (Ayanaba and Dart, 1977).

1.6.6 **Fertilizer use**

Since bambara groundnut is grown as a low-input subsistence crop, recommendations for optimal fertilizer requirements are scarce. Superphosphate at planting applied at the rate of 60 kg ha^{-1} and sulphate of ammonia at 40 kg ha^{-1} three weeks after sowing have been recommended (Stanton *et al.*, 1966). Where the crop is grown after a well-fertilized species, such as maize, on intensively managed soils, a maintenance dressing with superphosphate (19% P_2O_5) at 112–170 kg ha^{-1} is advocated at or after planting. On less fertile soils a compound fertilizer at 225–450 kg ha^{-1} is recommended as an initial application (Johnson, 1968). However, experiments in Nigeria, Malawi and Swaziland have generally failed to show any responses of bambara groundnut to the application of single superphosphate at rates of up to 336 kg ha^{-1} (Nyasaland Protectorate, 1963; Malawi Government, 1965, 1967, 1971; Cumberland, 1978; Nnadi *et al.*, 1981). In some cases phosphate fertilizer even appeared to decrease yields.

In a study of bambara groundnut cultivation at 25 locations in the Western Province of Zambia, Linnemann (1990) reported that no farmers had applied any fertilizer or manure in the 1989–90 season. Those who had tried to do so in the past had experienced adverse results.

Responses to nitrogen are generally thought to be poor. For example, a starter application rate of 20 kg ha^{-1} did not increase yields (Nnadi *et al.*, 1976) and an application of sulphate of ammonia at 112 kg ha^{-1} had only a marginal effect on yields (Malawi Government, 1971) or even depressed them (Malawi Government, 1970).

In Zaire, seed yields increased from 694 to 843 kg ha^{-1} after application of lime at 500 kg ha^{-1} (Belgium Government, 1960a).

An application of 25 kg ha^{-1} of potassium increased yields in an experiment in Swaziland (Swaziland Government, 1979).

When fertilizer is applied at planting, it should be broadcast prior to the final harrowing or ridging on heavy soils. On light soils, it is preferable to band it before planting (Johnson, 1968).

1.6.7 **Weeding and ridging**

Depending on the extent of weed infestation, farmers may weed between one and three times, usually by hoe. Weeding is often combined with ridging or 'earthing up' in order to cover the pegs so that pods can penetrate the soil surface (Holm and Marloth, 1940; FAO, 1961; Johnson, 1968). To minimize damage to the establishing pegs, it is important that ridging does not occur too close to the crop rows or when the soil is wet.

The results from experiments on ridging are contradictory, with reports of both increases and reductions in yield (Malawi Government, 1977; Malawi ARC 1975a). In general, yields were significantly depressed when the earthing up was done during the appearance of second flowers. Yield reductions have also been accompanied by increased incidence of disease (van der Wolk, 1915; Malawi ARC, 1972).

1.6.8 **Pests and diseases**

The crop appears to be singularly free of diseases (Johnson, 1968) and has a reputation for resisting pests. It is less susceptible to diseases than groundnut, possibly because it is often grown in small-scale mixed intercropping with other similarly isolated plants, and as a result of the wide genetic diversity cultivated.

Fungi

Table 1.7 summarizes the fungi that attack bambara groundnut. Leaf spot (*Cercospora*) and mildew (*Erysiphe polygoni*) are reported to damage the crop in Swaziland. Lamptey and Ofei (1977) investigated the effect of (*Cercospora*) on three accessions of bambara groundnut and found that the number of flowers and pods decreased significantly when the disease occurred before flowering. Plants inoculated before flowering also had small pods. No significant yield differences were obtained between plants inoculated at various stages of flowering and the control. These observations confirmed the earlier observations of Teyegaga (1970).

Leaf spot caused by *Cercospora canescens* is usually more severe after late planting (Billington, 1970). Planting resistant genotypes is the best method to prevent this fungal disease (Kinyawa, 1969; Kannaiyan, 1988).

Cumberland (1978) conducted an experiment in which the insecticide carbaryl (Sevin) and the fungicide chlorothalonil (Daconil) were used alone and in combination to control, among others, *Erysiphe polygoni*. Control of pests and diseases was good but no significant yield differences were found between the treatments and the control. In contrast, Haque (1980) reported a 50% increase in seed yield after weekly treatment with 110 g carbaryl and 36 g chlorothalonil ha^{-1}. In Malawi, a fortnightly application of a copper

Table 1.7 Fungi reported to attack bambara groundnut

Ascochyta phaseolorum	Bouriquet, 1946
Cercospora canescens	Nyasaland, 1960; Wallace and Wallace, 1947
Cercospora voandzeiae	Bouriquet, 1946
Colletotrichum capisci	Hepper, 1970
Corticium solani (*Rhizoctonia solani*)	Deighton, 1956; Leather, 1959; Malawi, 1972
Elsinoe spp.	Zambia, 1975
Erysiphe polygoni	Haque, 1980
Fusarium oxysporum f. sp. *voandzeiae*	Armstrong *et al.*, 1975
Leptospaerulina trifolii	Hepper, 1970
Meliola vignae-gracilis	Deighton, 1956
Phaeolus manihotis	Bouriquet, 1946
Phomopsis sp.	Kannaiyan, 1988
Phyllosticta voandzeiae	Deighton, 1956; Leather, 1959
Sclerotium rolfsii (*Corticium rolfsii*)	Maduewesi, 1975
Sphaerotheca voandzeiae	Bouriquet, 1946
Synchytrium dolichi	Hepper, 1970

fungicide, Kocide 101, against mildew increased yields, though differences were not significant (Malawi ARC, 1975b).

Soil type, drought and nematode infestation influence the incidence of the wilting disease caused by *Fusarium oxysporum* (Armstrong *et al.*, 1975). Resistant genotypes are the best answer (Kinyawa, 1969); populations planted in Tanzania showed considerable variation in resistance (Ebbels and Billington, 1972).

Virus

Table 1.8 summarizes the viruses that have been recorded on bambara groundnut. Vectors of viruses include *Medythia quaterna*, which carried cowpea mottle virus from cowpea to bambara groundnut (Allen *et al.*, 1981), and *Bemisia tabaci*, which transmitted horsegram yellow mosaic virus (Muniyappa and Veeresh, 1984).

Most of the bambara groundnut accessions held at IITA were found to be susceptible to cowpea aphid-borne mosaic virus (Ng *et al.*, 1985). A virus occurring in bambara groundnut in northwestern and eastern Tanzania was found to be serologically related to peanut mottle virus (Bock, 1973; Bock *et al.*, 1978). Although the bambara isolate did not infect groundnut, it did infect *Chenopodium amaranticolor* Coste and Reyn., a species which was formerly reported as immune to peanut mottle and thus considered to be of diagnostic value.

Table 1.8 Viruses reported to attack bambara groundnut

Alfalfa (lucerne) mosaic	Klesser, 1961a
Alsike clover mosaic	*idem*
Bean chlorotic ringspot	Klesser, 1961b
Bean local chlorosis	*idem*
Bean mosaic	*idem*
Bean necrosis	*idem*
Cowpea aphid-borne mosaic	Ng *et al.*, 1985
Cowpea mottle	Allen, 1980
Horsegram (Dolichos) yellow mosaic	Muniyappa and Veeresh, 1984
Voandzeia necrotic mosaic	Monsarrat *et al.*, 1984
White clover mosaic	Klesser, 1961a

Parasites

An angiospermous hemiparasite, *Alectra vogelii* Benth., is reported to live on the roots of bambara groundnut. The Swazi name for the parasite is 'Mlilwane' which means 'little fire'. This reflects the appearance of an infested field; eventually the field will look as if it has been burnt (Beck, 1987). The seed of the parasite is very small. Food reserves are quickly depleted and the plant therefore adopts a holoparasitic habit until the green shoot emerges above the ground. From this stage onwards the parasite's habit becomes hemiparasitic (Visser *et al.*, 1977). *Alectra vogelii* also parasitizes the roots of certain other legumes such as groundnut and cowpea, as well as the roots of some non-leguminous crops, such as sunflower (*Helianthus annus* L.) (Visser, 1975). In Swaziland, yields were halved as a result of infestation by the parasite. The use of guar (*Cyamopsis tetragonoloba* (L.) Taub.) as a catch crop seemed promising (Beck, 1987). The germination, growth, development and metabolism of the parasite have been studied in detail (Visser, 1975; Dorr *et al.*, 1977; Visser *et al.*, 1977; Dorr *et al.*, 1979; Gouws *et al.*, 1980).

Nematodes

Root-knot nematodes *Meloidogyne javancia* and *Meloidogyne incognita*, can seriously affect the yield of bambara groundnut (Martin, 1959; Johnson, 1968; Ogbuji, 1979). About ten weeks after planting, the leaves turn yellowish and the plants become stunted and die. The level of nematode infestation is reportedly much greater on light-textured soils (Johnson, 1968).

Pests

Linnemann (1990) reports that germinating seeds of bambara groundnut occasionally suffer damage from rats, termites, red ants, or cutworms (probably *Agrotis* spp.) which sever the seedlings at ground level. Pests such as duikers (*Sylvicapra grimmia*) and grasshoppers may feed on leaves of established plants. Insect pests such as *Ootheca mutabilis* and other foliage feeders, such as the leaf hoppers *Empoasca facialis* and *Hilda patruellis*, can also cause damage (Rassel, 1960; Raheja, 1976). The larvae of *Diacrisia maculosa* and *Lamprosema indicata* are also field pests (Hepper, 1970). The American bollworm (*Heliothis armigera*) is one of the most serious pests in Swaziland (Haque, 1980).

In Western Zambia, maturing seeds may be attacked by rats, brown ants, and even monkeys, wild pigs and bushbabies (*Galago* sp.) (Linnemann, 1990). Developing pods may also be injured by *Piezotrachelus ugandum* (Marshall, 1936). Oviposition takes place when the pods are well-formed and beginning to harden. At harvest the interior of the pods may be completely reduced to a brown faecal mass. Farmers in Tanzania reported that the pest is worst in years of heavy rainfall.

Larvae of the genus *Rivellia* cause damage to root nodules of the crop. In the region of Yangambi in Zaire the infestation was not related to a definite period of the year owing to the uniformity of the climate. No influence of the soil or previous crops was found. High rates of infestation cause an imbalance in the nitrogen nutrition of the plant (Seeger and Maldague, 1960).

1.6.9 **Harvesting**

Harvested seeds may be eaten either immature or fully ripe. Immature seeds are usually eaten fresh (and are more palatable than more mature harder seeds). Small plots may be harvested sequentially as seeds are required for consumption. The practice of harvesting the early-maturing plants first and using these for consumption implies a continuous selection of late-maturing types for subsequent sowing. This may be an undesirable trait under rainfed conditions (Figure 1.5). Linnemann (1990) reported that in the Western Province of Zambia farmers generally thresh pods in a pestle and mortar and then select the larger, undamaged seeds for subsequent planting. In contrast, some farmers selected their planting material immediately after harvest and stored it unthreshed and separate from the seed allocated for consumption. On larger fields the crop may be harvested during a dry period when the leaves rapidly senesce. Plants are left in the field to dry completely and are then collected in one harvest.

The leaves and haulms may be used as animal feed and analysis of dried leaves show crude protein 15.9%, crude fibre 31.7%, ash, 7.5% and fat 1.8% (Kay, 1979).

Although crop duration is extremely variable, harvesting usually takes place between 90 and 170 days after sowing, depending on genotype, temperature and other environmental factors (Johnson, 1968; Doku and Karikari, 1970). Plants are usually uprooted by hand or by using a hoe.

Figure 1.5 A bambara groundnut field in which the early maturing plants have been harvested for consumption; the late-maturing plants will provide next year's seed.

Occasionally, a single-furrow ox-drawn plough is used. Since many pods are detached from the haulm during harvesting, gleaning is important if maximum yield is to be obtained. Field losses in this way as high as 50–65% of total yield have been reported (Nyasaland Protectorate, 1959; Johnson, 1968).

After drying in the sun to about 12% moisture content, pods are stored or shelled by hand with flails or using modified groundnut shellers (Kay, 1979). Typically, shelling percentages range between about 70 and 77% by weight of pods (Johnson, 1968; Babiker, 1989). However, under conditions of extreme water stress, shelling percentages may be as low as 50% for a crop which achieved a pod yield of only 40 kg ha^{-1}. After shelling seeds are stored in plastic bags to prevent insect attack.

1.6.10 **Yield and production**

As to be expected with primitive cultivars, Bambara groundnut yields vary considerably between sites, seasons and genotypes. The highest recorded seed yields in field conditions are close to 4000 kg ha^{-1} (See Section 1.8). However, in most circumstances yields are considerably lower. For example, yields as low as 60–110 kg ha^{-1} have been reported in Zambia and for most of the semi-arid tropics typical yields are often between 650 and 850 kg ha^{-1} (Stanton *et al.*, 1966).

Rainfed crops in dry regions of the tropics are normally vulnerable to large yield fluctuations from one year to the next. Rather than presenting absolute yield values it is therefore more useful to emphasize that bambara groundnut produces reasonable yields under drought conditions on poor, pest- and disease-ridden soils where the cultivation of similar crops such as groundnut, common bean and cowpea is too risky. (Ezueh, 1977; Rachie and Roberts, 1974). Bambara groundnut clearly has the potential both for high yields under optimal conditions and for stable yields under drought conditions.

1.7. POST-HARVEST TECHNOLOGY

1.7.1 **Storage**

Traditionally, the crop is shelled by hand, using a flat stone or pestle and mortar. This can be a tedious operation because the shells closely fit the kernels. The optimum stage for shelling is when the pods rattle on shaking. If shelled seeds have to be stored, for future sowing, they should ideally be treated with 0.27 kg of 5% insecticidal dust per 100 kg seed, preferably after fumigation with phosphine-producing tablets or carbon disulphide (Johnson, 1968).

The insects that attack bambara groundnut seeds infest the crop in the field and are then carried to the store. The resulting damage produces seeds of reduced weight, viability and marketability. The worst storage pests of the crop are *Callosobruchus maculatus* (Ofuya and Bamigbola, 1991) and *C. subinnotatus* (Southgate, 1958; Prevett, 1967; 1970; Caswell, 1980; Amuti and Larbi, 1981).

Pereira (1983) studied the effectiveness of six vegetable oils as protectants against infestation by *Callosobruchus maculatus* in stored cowpea and bambara groundnut seeds. He found that, whereas in cowpea only neem kernel oil reduced oviposition, in bambara groundnut, neem kernel oil, palm oil, groundnut oil and karite oil all reduced oviposition. Furthermore, neem oil at 5 ml kg^{-1} protected bambara groundnut seeds for up to 6 months. The protective action of neem oil was ascribed to a combination of ovicidal and larvicidal activity as well as reduced oviposition and it was suggested that the smooth, hard bambara groundnut testa offered a better surface for a uniform coating of oil than the more wrinkled cowpea seeds.

1.7.2 **Processing**

At present, the commercial potential of bambara groundnut remains largely unexploited. Some canning has been reported in Ghana and Zimbabwe.

Poulter and Caygill (1980) found that the processing of vegetable milk from bambara groundnut seeds is also feasible and homogenization of the whole seed flour in water together with coconut or soya bean oils yielded a stable emulsion which could be maintained after pasteurization. They also found that curds, similar to those prepared from soya bean, could easily be prepared from bambara groundnut milk. The bland, creamy-white product was high in protein and had a trypsin inhibitor content which was very much lower than that of the untreated vegetable milk.

In Ghana a traditional maize food ('ablemamu') which has a low protein content, has been successfully supplemented with bambara groundnut flour to provide an acceptable and tasty product called 'bamcorn' (Lartey, 1976).

Similarly, bambara groundnut seeds have been successfully mixed with sorghum in a ratio of 70:30 in order to improve the protein quality of feed (Obizoba, 1989). In this study, the effects of soaking and dehulling cream- and brown-coloured seeds of bambara groundnut were compared in feeding trials with rats. Dehulling significantly improved the nutritive value of the cream-seeded accession by more than that of the brown-seeded accession, whilst soaking had the converse effect. The feed intake of the cream-seeded accession was greater than that of the brown-seeded accession and was significantly greater in all the bambara groundnut treatments than the casein control diet. The results of this study indicate that processing to produce blends of bambara groundnut and sorghum could provide cheap and nutritious diets for people dependent on the crops as staple foods.

Obizoba (1983) reported that, in general, dishes prepared from cowpea and bambara groundnut seeds are mainly consumed by older children and adults. Because of their limited digestive capacities, infants and weaning children are thought to be unable to eat these products. However, when cowpea or bambara groundnut flour was supplemented with rice it formed a good quality protein food which was acceptable to infants and small children.

A major reason for avoiding legumes, particularly as an infant food, is the difficulty experienced in digesting them. The consumption of pulses is commonly associated with flatulence, abdominal pain and diarrhoea. The production of flatus by monogastric animals is due to the colonic fermentation of carbohydrates which escape breakdown in the stomach and small intestine. The oligosaccharides raffinose, stachyose and verbascose are thought to be the major producers of flatus hydrogen. Phillips and Abbey (1989) investigated the composition and flatulence-producing potential of a number of legume species. In a comparison of 16 varieties of 8 species they reported that only bambara groundnut produced no more flatus hydrogen that the control diet. This evidence appears to refute the conclusion of Obizoba (1983) that bambara groundnut by itself is unsuitable as an infant food.

1.8 GENETICS AND BREEDING

Local selection in bambara groundnut under cultivation has led to a gradual increase in pod and seed size and a general shortening of internode length resulting in a bunched habit (Smartt, 1985); this form tends to be routinely selected because it facilitates harvesting. Another factor selected for is faster, more uniform germination in today's cultivated types (Hepper, 1970). However, in West Africa, wild and domesticated populations appear to intercross, thus eroding the effects of selection (Smartt, 1985).

Diverse genotypes, and their associated characteristics such as disease resistance and drought tolerance, are recognized by local farmers, who use the appearance of the seed coat as a means of identification (Dalziel, 1937 and Johnson, 1968). Criteria useful for plant breeding, according to Tardieu (1958), are: habit (tall or short), flower colour (yellow, green or orange), pod size (range 5–15 mm), pod thickness (thick or thin), pod surface (wrinkled, hammered or smooth), pod colour (white, orange-yellow or purple) and number of seeds per pod. To date, breeding of bambara groundnut has focused on selection between and within population samples for yield performance, disease resistance and drought tolerance. The best results from trials aimed at yield performance are presented in Table 1.10. The highest seed yield ever recorded was 3870 kg ha^{-1} in Zimbabwe (Johnson, 1968). Search for wide adaptability is needed to avoid the current restrictions on using only local materials in the particular areas. Selection for disease resistance in bambara groundnut has been done only on a limited scale. *Fusarium* wilt and leaf spot caused by the fungus *Cercospora canescens* are two important diseases for which differences in resistance between genotypes exist (Kinyawa, 1969). The Ukirigur bambara improvement programme was, in part, concerned with selection for resistance to *Fusarium oxysporum* (Ebbels and Billington, 1972).

In an important study in which germplasm was screened, drought-tolerant genotypes were identified in the germplasm collection at IITA (Begemann, 1988). The Cameroonian accessions TVsu 395, TVsu 438 and TVsu 501 escaped drought by maturing early and a deep root system enabled TVsu 674, TVsu 942, TVsu 1153 and TVsu 1207 to yield well under conditions of limited rainfall.

It is also important to characterize primitive cultivars through a descriptor system. Doku (1969) used the ratio of petiole length to length of the fourth internode from the main axis to classify local forms into three groups: bunched (8.2–11.0), semi-bunched (7.4–7.7) and spreading (4.4–6.5). Later Doku and Opoku-Asiama (1978) described new criteria to identify habit using the number of branches, plant base spread, canopy spread, leaflet length and width, and petiole and internode length. Ezedinma and Maneke (1985) used yet another set of criteria, based on seed colour and canopy diameter at 100 days after planting. Forms with a canopy diameter less than 40 cm were classified as bunched, between 40 and 80 cm as semi-bunched and greater than 80 cm as spreading.

Certain morphological traits, such as internode length, plant base spread, canopy spread and leaflet size, are not highly heritable, thus restricting their use in classification to well-defined circumstances. For instance, leaflet size is influenced by soil fertility, whereas plants grown at a day length of 14 h have longer internodes than plants grown at a photoperiod of 12 h (Linnemann, 1991a). Even testa colour is not always a reliable descriptor for a genotype; several instances of resegregating populations have been recorded (for example Mittelholzer, 1969; Linnemann, 1990).

A comprehensive descriptor list for bambara groundnut, published by the International Board for Plant Genetic Resources, was developed by F. Begemann and Q. Ng (IBPGR/IITA/GTZ, 1987) to facilitate the systematic description of germplasm accessions in collections and to facilitate the exchange of accessions between institutions. Accessions are characterized by recording highly heritable traits which are visible with the naked eye and expressed in all environments (Figure 1.6). It is important to describe

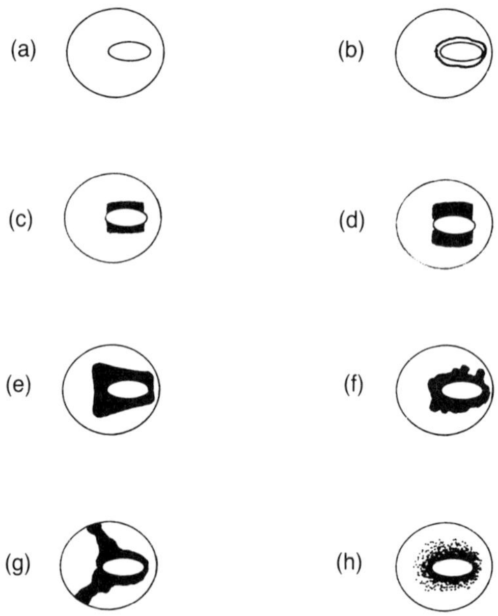

Figure 1.6 Eye patterns of bambara groundnut seeds (a) no eye; (b) eye as a circle around hilum; (c) eye as two thin lines on both sides of hilum; (d) eye as two thick lines on both sides of hilum; (e) eye shape almost triangular; (f) large eye, more or less shapeless, becoming frayed; (g) eye like a butterfly; (h) mottled eye. Adapted from IBPGR/IITA/GTZ (1987).

the seed accurately in terms of attributes such as testa colour, testa pattern, colour of pattern, eye pattern and eye colour.

The largest germplasm collection of about 2000 accessions, is held by the International Institute of Tropical Agriculture in Nigeria (IITA, 1990). Smaller collections are maintained by the Institut d'Economie Rurale (SRCVO) in Bamako, Mali, the Msekera Regional Research Station in Chipata, Zambia (Kannaiyan, 1988), the Faculty of Agriculture, Forestry and Veterinary Science of the University of Dar-es-Salaam, Tanzania (Munene, 1982), the Department of Crop Science of the University of Ghana in Legon (Doku *et al.*, 1978) and the National Bureau of Plant Genetic Resources in New Delhi, India (Chandel and Singh, 1984). The most recent collecting work, with support of the International Board for Plant Genetic Resources and IITA, acquired accessions from Botswana (Appa Rao *et al.*, 1986), Ghana (Ayensu, 1989), Sudan (Croston and El Ahmadi, 1982), Swaziland (Padulosi *et al.*, 1989), Tanzania (Padulosi and Ng, 1989b), Togo (Mergeai, 1986), Zambia (Mehra, 1982) and Zimbabwe (Padulosi and Ng, 1989a). According to Begemann (1988) the existing germplasm collections hold insufficient population samples from Chad, Ethiopia, Niger and Sudan. Particular attention needs to be paid to collecting from high rainfall areas.

1.9 TRADE

Production, consumption and sale of bambara groundnut are mostly for local requirements, hence production statistics are not readily available. However, of the leguminous crops in Africa bambara groundnut ranks third in production after groundnut and cowpea (Sellschop, 1962). Estimated world production is 330 000 ton annually, of which about half is produced in West Africa (Coudert, 1982). Table 1.9 presents the estimated annual production of bambara groundnut for a number of African countries.

Table 1.9 Estimated annual production of bambara groundnut in some African countries

Country	*Production (t year^{-1})*	*Reference*
Nigeria	50 000	Coudert, 1984
Mali	35 000	*idem*
Mozambique	24 000	Heemskerk *et al.*, 1987
Chad	15 000–20 000	Coudert, 1984
Ghana	15 000–20 000	*idem*
Niger	12 000	*idem*
Togo	8 000	*idem*
Benin	4 000	*idem*
Swaziland	1 500	Swaziland, 1983

At the end of the 19th and the beginning of the 20th century small quantities of bambara groundnut were exported to Western Europe, mainly for cattle food (Zagorodsky, 1911; Hepper, 1970). Trade within West Africa is traditionally from countries bordering the Sahara southwards to coastal markets. Initially these trade connections followed the caravan routes. Nowadays railways and roads provide the necessary routing. The main exporting countries are Niger, Chad, Burkina Faso, Mali and Senegal, which supply markets in Nigeria, Benin, Togo and Ghana. The export volume is difficult to monitor, as trade in bambara groundnut consists of a multitude of small quantities taken across national borders by traders or producers themselves. Along the frontiers of Niger, Benin and Nigeria border trade is particularly active (Coudert, 1984).

According to Coudert (1982), the average market price for bambara groundnut in West Africa lies well above the world price of other pulses, indicating that the quantities offered for sale are insufficient to satisfy demand. Producers could be stimulated to increase output and exporting countries could benefit by increased earnings. At the same time, importing countries would benefit from an improved food supply. Promotion of regional trade will require better transport between countries; better information for exporters, importers, producers and consumers about prices and supplies would improve trade and reduce speculation.

1.10 FARMERS' PERCEPTION OF THE CROP

Bambara groundnut is an important secondary food crop in rainfed agriculture in the drier parts of Africa. The crop is usually cultivated on small farms by the poorer section of the population, often by women. In West Niger for instance, bambara groundnut is also known as 'the groundnut of the women' (Coudert, 1982). On part of the acreage that was traditionally cultivated with bambara groundnut, groundnut is now grown mainly because there is more demand on domestic and international markets.

An important trait of bambara groundnut is the drought tolerance of many genotypes. Farmers claim that in years when the groundnut crop fails due to bad rains, their bambara groundnut still gives them some returns (Linnemann, 1990).

Farmers also report that the crop is relatively free of pests and diseases. This is not necessarily a genetic characteristic; it could be largely the result of intercropping and the heterogeneous composition of most seed lots. In northern Nigeria, for instance, the crops were indeed remarkably free of pests and diseases (Linnemann, 1988), but in Western Province, Zambia, serious losses caused by *Fusarium* wilt were recorded (Linnemann, 1990).

Another reason farmers appreciate bambara groundnut is because it tolerates poor soils.

Figure 1.7 Selling bambara groundnut cooked in their pods, on the pavement in Livingstone, Zambia.

Farmers also like bambara groundnut for its diverse uses. But, primarily, it is often preferred for its taste and flavour when compared with groundnut. Freshly harvested pods are cooked, shelled and eaten as a snack during the later part of the rainy season, thus providing a source of food before most of the other crops have been harvested (Figure 1.7).

During the dry season the dried seeds are cooked and eaten as a snack or in relish with the evening meal. Bambara groundnut is also used as a main dish, for instance by cooking with nibbed maize or by preparing a porridge solely from bambara groundnut flour. Travellers remark that they are sustained for long periods by a meal of bambara groundnut.

Finally, bambara groundnut may be used as a source of additional income. Any excess production can easily be sold along the road as a snack to passers-by or in the local markets. Many farmers do not want to totally oust bambara groundnut by groundnut, because they consider the two crops complementary both in cultivation and use (Linnemann, 1990).

1.11 FUTURE POTENTIAL

Experimental evidence and the more recent research indicate that bambara groundnut is a promising crop which has been neglected by international

Table 1.10 Highest seed yields from bambara groundnut trials in some African countries

Country	*Yield (kg ha^{-1})*	*Germplasm accession*	*Reference*
Madagascar	2100	Rangovava	Dufournet, 1957
Malawi	3784	Barotseland	Nyasaland, 1960
Mozambique	1378	INIA 37	Heemskerk *et al.*, 1987
Nigeria	1339	TVsu 1034	Goli and Ng, 1988
Swaziland	1262	Brown speckled	Haque, 1980
Tanzania	1809	Kasholishoro	Smyth, 1968
Zaire	1792	Unknown	Belgium, 1960b
Zambia	2027	FAO 142/80	Kannaiyan, 1988

agricultural science. The crop combines the potential to produce substantial yields under favourable growing conditions (see Table 1.10) with the ability to produce a small but stable yield under marginal conditions of limited rainfall. In many dry, hostile environments bambara groundnut is able to produce at least some yield in circumstances in which groundnut and other, drought-tolerant legumes may fail completely. In addition to being an ideal 'food-security' crop, bambara groundnut contributes to the sustainability of the cropping system. As a legume, it is useful in crop rotation as it provides a source of nitrogen for the following crop. Furthermore, because it is relatively free of pests and diseases its inclusion in the cropping sequence may delay their build-up in succeeding crops in the rotation.

An important point, which is often ignored by potential sponsors of research, is that many Africans prefer bambara groundnut to other legumes for its unique taste and flavour. The crop has a long tradition within indigenous African cultures as a locally available crop (Table 1.10). Human movement from rural areas to urban centres has resulted in a vast internal market within Africa which has hardly been tapped. Evidence indicates that the price of bambara groundnut in city markets is higher than that of other legumes (Coudert, 1984; Linnemann, 1990) and this is often because demand exceeds supply.

The improvement of a crop species involves the manipulation of its genetic potential through plant breeding and the amelioration of its production environment through agronomic management. Thus, as a prerequisite, an integrated programme of plant breeding is essential in order to assess more fully and improve the genetic potential of bambara groundnut. This would best be done by using the concept of a crop *ideotype*. In this way specific ideotypes can be defined for different environments and production objectives, such as yield, quality and sustainability.

For high rainfall and/or commercial production systems, appropriate ideotypes would be selected for high yield potential and quality aspects, such as seed colour and size and cooking and nutritional factors. In terms of crop morphology, individual plants would have larger leaves than is typical of primitive cultivars. Thus, the crop canopy would achieve complete ground cover earlier in the season, allowing the maximum amount of solar radiation to be usefully intercepted by green foliage.

By contrast, an ideotype for rainfed, subsistence systems would have many characteristics similar to current primitive cultivars. The ability to withstand prolonged periods of water stress and resistance to pests and diseases would be more important than yield potential *per se*. A 'subsistence' ideotype for arid environments might have a high root/shoot ratio, i.e. a greater proportion of assimilates would be allocated to roots, with a relatively small, compact leaf canopy. Any breeding programme will require the expansion and fuller use of existing germplasm collections.

In addition to the development of improved genetic material, research should be conducted to improve the agronomy of the crop. This research should identify optimal management practices in both sole and intercrop systems as well as for commercial and subsistence production.

Much previous work on the agronomy of bambara groundnut has been on a short-term and trial-and-error basis. This has not been cost-effective either in terms of time or in terms of the very limited amount of funds channelled to research on underutilized species. An alternative to piecemeal and sporadic research efforts is to develop robust crop simulation models to identify the agroclimatic regions suitable for the production of bambara groundnut.

Besides having great potential for Africa, there are possibilities for expanding bambara groundnut production to other semi-arid areas of the world. The application of simulation models on detailed ecophysiological data, as for instance demonstrated by Hornetz (1988) for dryland in South-east Kenya, could help to define potential production areas. Models could also be used to predict likely yields and economic and/or nutritional returns in comparison with other drought-tolerant legumes and therefore avoid the expense and effort involved in repetitive multi-location trials. Another important role for models would be the identification of research priorities and objectives so that effort could initially be concentrated on the most important aspects of the crop.

Finally, the possibility for bambara groundnut as a cash crop for processing and world trade deserves more consideration. Establishing a niche for the crop in terms of food processing, confectionery or snack products would help to provide the impetus for bambara groundnut becoming a more important world crop.

REFERENCES

Allen, D.J. (1980) Identification of resistance to cowpea mottle virus. *Trop. Agriculture*, **57**, 325–32.

Allen, D.J., Anno-Nyako, F.O., Ochieng, R.S. and Markandu Ratinam (1981) Beetle transmission of cowpea mottle and southern bean mosaic viruses in West Africa. *Trop. Agriculture*, **58**, 171–5.

Ameyaw, C.E.G. and Doku, E.V. (1983) Effects of soil moisture stress on the reproductive efficiency and yield of the bambara groundnut (*Voandzeia subterranea*). *Tropical Grain Legume Bull.*, **28**, 23–9.

Amuti, K.S. and Larbi, M. (1981) Postharvest losses in the bambara and geocarpa groundnut seeds stored under traditional conditions in Ghana. *Tropical Grain Legume Bull.*, **23**, 20–2.

Amuti, K.S. and Pollard, C.J. (1977a) Soluble carbohydrates of dry and developing seeds. *Phytochemistry*, **16**, 529–32.

Amuti, K.S. and Pollard, C.J. (1977b) The metabolism of galactose and the raffinose oligosaccharides by germinating bambara groundnut seeds. *Phytochemistry*, **16**, 533–7.

Appa Rao, S., Mazhani, L.M. and Attere, A.F. (1986) Collecting in Botswana.*Plant Genetic Resources Newsletter*, **68**, 27–8.

Armstrong, G.M., Armstrong, J.K. and Billington, R.V. (1975) *Fusarium oxysporum* forma specialis *voandzeiae*, a new form of species causing wilt of bambara groundnut. *Mycologia*, **67**, 709–14.

Ayanaba, A. and Dart, P.J. (1977) *Biological Nitrogen Fixation in Farming Systems of the Tropics*, John Wiley, Chichester.

Ayensu, F.K. (1989) Bambara groundnut collecting in Ghana. *Plant Genetic Resources Newsletter*, **77**, 42.

Babiker, A.H. (1989) *The Growth, Dry Matter and Yield of Bambara Groundnut (*Vigna subterranea *(L.) Verd.) and Groundnut (*Arachis hypogaea *(L.)) in Response to Irrigation and Drought*, M.Sc. Thesis, University of Nottingham.

Beck, B.D.A. (1987) The effect of attack by *Alectra vogelii* (Benth.) on the yield of njugo beans *Vigna subterranea* (L.) Verdc., in *Parasitic Flowering Plants*, (eds H. Chr. Weber and W. Forstrenter), Proceedings of the 4th International Symposium on Parasitic Flowering Plants, Marburg 1987, Marburg, pp. 79–82.

Begemann, F. (1988) *Ecogeographic Differentiation of Bambarra Groundnut (*Vigna subterranea*) in the Collection of the International Institute of Tropical Agriculture (I.I.T.A.)*. Wissenschaftlicher Fachverlag, Giessen, Germany.

Belgium Government (1960a) *Jaarverslag voor het dienstjaar 1958*, Nationaal Institut voor Landbouwstudie in Belgisch-Congo. Clarence Denis, Brussels.

Belgium Government (1960b) *Rapport annuel pour l'exercise 1959*, Institut National pour l'Etude Agronomique du Congo Belge, Clarence Denis, Brussels.

Billington, R.V. (1970) The effect of planting dates on disease incidence on legumes and its effect on final yields. *Ukiriguru Research Notes*, **44**, 5–7.

Bock, K.R. (1973) Peanut mottle virus in East Africa. *Annals Appl. Biol.*, **74**, 171–9.

Bock, K.R., Guthrie, E.J. and Meredith, G. (1978) Viruses occurring in East Africa that are related to peanut mottle virus. *Annals Appl. Biol.*, **89**, 423–8.

Bouriquet, G. (1946) Les maladies des plantes cultivées à Madagascar; maladies du *Voandzeia subterranea*. *Encyclopedie Mycologique*, **12**, 352–9.

Brook, C., Bray, J., Cumberland, J. *et al.* (1977) Field crop, horticultural and pasture recommendations for Swaziland: jugo beans (bambarra groundnuts, tindlubu). *Advisory Bulletin No. 1*, Agricultural Research Division, U.C.S. Faculty of Agriculture, Malkerns Research Station.

Brooks, C.B., Dadson, R.B. and Green, B.M. (1988) Evaluation of symbiotic effectiveness

of elite and wild strains of *Bradyrhizobium* on cultivars of *Voandzeia subterranea* (L.) Thouars. *Tropical Agriculture*, **65**, 61–3.
Brough, S.H. and Azam-Ali, S.N. (1992) The effect of soil moisture on the proximate composition of bambara groundnut (*Vigna subterranea* (L.) Verdc.). *Journal of the Science of Food and Agriculture*. (submitted)
Brough, S.H. (1991) Biochemical composition of bambara groundnut (*Vigna subterranea* (L.) Verdc.) in relation to soil moisture (personal communication).
Caswell, G.H. (1980) A review of the work done in the entomology section of the Institute for Agricultural Research on the pests of stored grain. *Samaru Miscellaneous Paper*, **99**, 1–12.
Chandel, K.P.S. and Singh, B.M. (1984) Some of our under-utilized plants. *Indian Farming*, **34**, 23–9.
Coudert, M.J. (1982) *Niébé et voandzou; une perspective pour le développement du commerce régional en Afrique de l'ouest*. FAO microfiche XF 8332635. International Trade Centre UNCTAD/GATT, Geneva.
Coudert, M.J. (1984) Market openings in West Africa for cowpeas and bambara groundnuts. *International Trade Forum*, **20**, 14, 15, 28, 29.
Croston, R.P. and El Ahmadi, A.B. (1982) Collecting in Darfur, Sudan. *Plant Genetic Resources Newsletter*, **50**, 8–31.
Cumberland, J. (1978) Jugo beans. *Research Report no. 19*, Agricultural Research Division, Ministry of Agriculture, Malkerns Research Station.
Dalziel, J.M. (1937) *Voandzeia* Thouars, in *The Useful Plants of West Tropical Africa*, Crown Agents, London, pp. 269–71.
Deighton, F.C. (1956) *Diseases of Cultivated and other Economic Plants in Sierra Leone*, Government Printing Department, Sierra Leone.
Dennis, E.A. (1977) Nodulation and nitrogen fixation in legumes in Ghana, in *Biological Nitrogen Fixation in Farming Systems in the Tropics*, (eds A. Ayanaba and P.J. Dart), John Wiley, Chichester, pp. 217–32.
Deshpande, S.S. and Damodaran, S. (1990) Food legumes: chemistry and technology. *Advances in Cereal Science and Technology* **10**, 147–241.
Doku, E.V. (1967) Are there any alternatives to the traditional bush fallow system of maintaining fertility? *Ghana Farmer*, **11**, 27–30, 38.
Doku, E.V. (1969) Growth habit and pod production in bambara groundnut (*Voandzeia subterranea*). *Ghana J. Agric. Science*, **2**, 91–5.
Doku, E.V. and Karikari, S.K. (1970) Fruit development in bambarra groundnut (*Voandzeia subterranea*). *Annals Bot.*, **34**, 951–6.
Doku, E.V. and Karikari, S.K. (1971a) Bambarra groundnut. *Economic Botany*, **25**, 225–62.
Doku, E.V. and Karikari, S.K. (1971b) The role of ants in pollination and pod production of bambarra groundnut. *Economic Botany*, **25**, 357–62.
Doku, E.V. and Opoku-Asiama, Y. (1978) A classification of bambara groundnut (*Voandzeia subterranea*) cultivars based on variation in morphological characters, in *Proceedings of a Symposium of Crop Improvement in Ghana*, Faculty of Agriculture, University of Ghana, Legon.
Doku, E.V., Hammonds, T.W. and Francis, B.J. (1978) On the composition of Ghanian bambara groundnuts (*Voandzeia subterranea* (L.) Thouars). *Tropical Science*, **20**, 263–9.
Dorr, I., Visser, J.H. and Albers, F. (1977) On the parasitism of *Alectra vogelii* Benth. (Scrophulariaceae) II. Origin of lateral roots in the contact area of the haustorium. *Zeit. Pflanzenphysiol.*, **85**, 349–60.
Dorr, I., Visser, J.H. and Kollmann, R. (1979) On the parasitism of *Alectra vogelii* Benth. (Scrophulariaceae) III. The occurrence of phloem between host and parasite. *Zeit. Pflanzenphysiol.*, **94**, 427–39.

Driessen, P.M. and Permadhy Sudewo (1977) A review of crops and crop performance on Southeast Asian lowland peats, in *Soil research; Final Report of the Agricultural Technical Assistance Programme 106 (1974–1977)*, Soil Research Institute, Bogor, pp. 48–9.

Dufournet, R. (1957) Note sur le voanjobory (*Voandzeia subterranea* Dup. Thouars) cultivé à Madagascar. *Riz et Riziculture et Cultures Vivriéres Tropicales*, **3**, 169–72.

Du Petit-Thouars, L.-M.A.A. (1806) *Genera Nova Madagascariensia*, Paris.

Ebbels, D.L. and Billington, R.V. (1972) *Fusarium* wilt of *Voandzeia subterranea* in Tanzania. *Trans. British Mycol. Soc.*, **58**, 336–8.

Elia, F.M. and Mwandemele, O.D. (1986) The effect of water deficit droughts on some plant characters in bambarra groundnut (*Vigna subterranea* Thouars). *Tropical Grain Legume Bull.*, **33**, 45–50.

Epias, L.G., DeFernandez, D.G. and Bressani, R. (1979) Possible effects of seed coat polyphenolics on the nutritional quality of bean protein. *J. Food Sci.*, **44**, 524–7.

Ezedinma, F.O.C. and Maneke, F.O. (1985) Preliminary studies on bambarra groundnut (*Voandzeia subterranea* Thouars) in the derived savanna belt of Nigeria. *Tropical Grain Legume Bull.*, **31**, 39–44.

Ezueh, M.I. (1977) Cultivation and utilization of minor food legumes in Nigeria. *Tropical Grain Legume Bull.*, **10**, 28–32.

FAO (1961) *Agricultural and Horticultural Seeds*. FAO Agricultural Studies no. 55, Food and Agriculture Organization, Rome, Italy.

FAO (1982) Legumes in human nutrition. *FAO Food and Nutrition Paper No. 20*, Food and Agriculture Organization, Rome, Italy.

Gaydou, E.M., Bianchini, J.-P. and Ratovohery, J.V. (1983) Triterpene alcohols, methyl sterols, sterols, and fatty acids in five Malagasy legume seed oils. *J. Agricul. Food Chem.*, **31**, 833–6.

Goli, A.E. and Ng, N.Q. (1988) Bambara groundnut multi-location yield trial. *Annual Report, Genetic Resources Unit, IITA*. 1987, 11–12.

Gouws, J., Visser, J.H. and Grobbelaar, N. (1980) Some aspects of the bidirectional translocation of 14C-labelled metabolites between *Alectra vogelii* Benth. and *Voandzeia subterranea* (L.) Thou. *Zeit. Pflanzenphysiol.*, **99**, 225–33.

Griffiths, D.W. (1979) The inhibition of digestive enzymes by extracts of field beans (*Vicia faba*). *J. Sci. of Food and Agric.*, **30**, 458–62.

Gueye, M. and Bordeleau, L.M., (1988) Nitrogen fixation in bambara groundnut, *Voandzeia subterranea* (L.) Thouars. *MIRCEN J.*, **4**, 365–75.

Guillemin, J.A., Perrottet, G.S. and Richard, A. (1832) *Florae Senegambiae Tentamen*, Paris.

Haque, I. (1980) Present status of grain legume research in Swaziland: 3. Jugo beans. *Tropical Grain Legume Bull.*, **17/18**, 38–40.

Harms, H. (1912) Über die Heimat der Erderbse, *Voandzeia subterranea* (L.) Thou. *Notizblatt des Königl. Bot. Gartens und Meseums zu Dahlen bei Steglitz (Berlin)*, **5**, 253–8.

Heemskerk, W., Reis Simango and Leonardo (1987), *Resultados de investigacao de segunda importancia 1982–1987*, INIA, Maputo.

Hepper, F.N. (1963) The bambara groundnut (*Voandzeia subterranea*) and Kersting's groundnut (*Kerstingiella geocarpa*) wild in West Africa. *Kew Bull.*, **16**, 395–407.

Hepper, F.N. (1970) Bambara groundnut (*Voandzeia subterranea*). *Field Crop Abstr.*, **23**, 1–6.

Holm, J.M. and Marloth, B.W. (1940) The bambara groundnut or njugo bean. *Farming in South Africa*, **15**, 195–8, 200.

Hornetz, B. (1988) Ecophysiological experiments for improving landuse patterns in the drylands of Southeast Kenya by means of drought resistant leguminous crops (tepary beans, bambarra groundnuts). *Tropenlandwirt*, **89**, 107–29.

IBPGR/IITA/GTZ (1987) *Descriptors for Bambara Groundnut*, International Board for Plant Genetic Resources, Rome.

International Institute of Tropical Agriculture (1990) *Annual Report 1989/90*, Ibadan.

Johnson, D.T. (1968) The bambara groundnut, a review. *Rhodesia Agric. J.*, **65**, 1–4.

Kannaiyan, J. (1988) *Review of research work on minor legumes*, Msekera Regional Research Station, Chipata.

Karikari, S.K. (1969) *Flowering, Pollination and Fruit Formation in Bambara Groundnut (*Voandzeia subterranea *Thouars) in Ghana*. M.Sc. Thesis, University of Ghana, Legon.

Karim, M. (1990) *The Growth, Yield, and Light Interception of Bambara Groundnut (*Vigna subterranea *(L.) Verdc.) and Groundnut (*Arachis hypogaea *(L.)) Grown under Different Irrigation Regimes*, M.Sc. Thesis, University of Nottingham.

Kay, D.E. (1979) *Food Legumes*. Crop and product digest no. 3, Tropical Products Institute, London.

Kinyawa, P.L. (1969) Diseases of bambara nuts at Ukiriguru. *Ukiriguru Research Notes*, 31.

Klesser, P.J. (1961a) The virus diseases of *Crotalaria, Glycine* and *Medicago* species. *Bothalia*, **7**, 497–519.

Klesser, P.J. (1961b) The virus disease of beans. *Bothalia*, **7**, 521–58.

Kutsch, H. and Schuh, H.-J. (1983) Simplified computer-based modeling of water balance in defined crop stands, in *Informationsverarbeitung Agrarwissenschaft*, (eds L. Reiner and H. Geidel), Badschurmpest in der Landwirtschaft, Stuttgart, **7**, 239–53.

Lamptey, P. and Ofei, I.O. (1977) The effect of leaf spot caused by *Cercospora canescens* (Ellis & Martin) on the yield of bambara groundnut (*Voandzeia subterranea* Thouars). *Ghana J. Agric. Science*, **10**, 113–15.

Lartey, B.L. (1976) The production of bamcorn, a maize–bambara groundnut product. *Ghana J. Agric. Science*, **9**, 139–42.

Leather, R.I. (1959) Diseases of economic plants in Ghana other than cacao. *Bull. Ghana Ministry of Food and Agriculture*, **1**, 5.

Lewitt, T. (1990) *Transpiration and Soil Surface Evaporation from Bambara Groundnut (*Vigna subterranea *(L.) Verdc.) and Groundnut (*Arachis hypogaea *(L.) Grown under Different Irrigation Regimes*, M.Sc. Thesis, University of Nottingham.

Linnaeus, C. (1763) *Species Plantarum*, 2nd ed, vol. 2, Laurentius Salvius, Holmiae.

Linnemann, A.R. (1987) Bambara groundnut (*Vigna subterranea* (L.) Verdc.) – a review. *Abstr. Trop. Agric.*, **12**, 9–25.

Linnemann, A.R. (1988) Cultivation of bambara groundnut (*Vigna subterranea* (L.) Verdc.) in Northern Nigeria. Report of a field study. *Tropical Crops Communication*, **15**, 14 pp.

Linnemann, A.R. (1990) Cultivation of bambara groundnut (*Vigna subterranea* (L.) Verdc.) in Western Province, Zambia. Report of a field study. *Tropical Crops Communication*, **16**, 19 pp.

Linnemann, A.R. (1991a) Preliminary observations on photoperiodic regulation of phenological development in bambara groundnut (*Vigna subterranea* (L.) Verdc.). *Field Crops Res.* (in press).

Linnemann, A.R. (1991b) Influence of photoperiod on development of bambara groundnut (*Vigna subterranea* (L.) Verdc.). *Acta Bot. Neerlandica* (in press).

Maduewesi, J.N.C. (1975) Host range and thermal inactivation of a cowpea isolate of *Sclerotium rolfsii*. *Nigerian J. Plant Protection*, **1**, 23–8.

Malawi Agricultural Research Council (1972) *Annual report for 1972*, Makoka Research Station, Thondwe, pp. 45, 47.

Malawi Agricultural Research Council (1975a) *Annual report for 1973*, Thondwe 40–41.

Malawi Agricultural Research Council (1975b) *Annual Report for 1974*, Thondwe, 33–34.

Malawi Government (1965) *Annual Report of the Department of Agriculture for the year 1962/63 (part II)*, Government Printer, Zomba.

Malawi Government (1967) *Annual Report of the Department of Agriculture for the year 1963/64 (part II)*, Government Printer, Zomba.

Malawi Government (1970) *Annual Report of the Department of Agriculture (part II) for the year 1964/65*, Government Printer, Zomba.

Malawi Government (1971) *Annual Report of the Department of Agriculture (part II) for the year 1966/67*, Government Printer, Zomba.

Malawi Government (1972) *Annual Report of the Department of Agricultural Research for the year 1963/64*, Government Printer, Zomba.

Malawi Government (1977) *Annual Report of the Department of Agricultural Research for the year 1973/74*, Government Printer, Zomba.

Marcgrav de Liebstad, G. (1648) *Historiae Rerum Naturalium Brasiliae*, Ioannes de Laet, Antwerp.

Maréchal, R., Mascherpa, J. and Stainier, F. (1978) Étude taxonomique d'un groupe complexe d'espèces des genres *Phaseolus* et *Vigna* (Papilionaceae) sur la base de données morphologiques et polliniques, traitées par l'analyse informatique. *Boissiera*, **28**, 177–8.

Marshall, G.A.K. (1936) Curculionidae (Col.) attacking cultivated plants. *Bull. Entomol. Research*, **27**, 253–9.

Martin, G.C. (1959) Plants attacked by root-knot nematodes in the Federation of Rhodesia and Nyasaland. *Rhodesia Agric. J.*, **56**, 162–75.

Martino-Ferrer, M. di and Ferrer, A. (1983) Trypsin inhibitor from bambara pea. *Phytochemistry*, **22**, 813–16.

Mehra, K.L. (1982) Collecting in Zambia. *Plant Genetic Resources Newsletter*, **50**, 45–50.

Mergeai, G. (1986) Bambara groundnut collecting in Togo. *Plant Genetic Resources Newsletter*, **67**, 38–40.

Mittelholzer, A.S. (1969) An evaluation of the yield of ten groundbean (*Voandzeia subterranea*) selections. *Annual Report on Grain Legume Research of the Agricultural Research Council of Malawi for October 1968 – September 1969*, Section 3.

Monsarrat, A, Fauquet, C. and J.-C. Thouvenel (1984) Une nouvelle maladie virale du pois de terre (*Voandzeia subterranea* Thouars) provoquée par un tymovirus, le Voandzeia Necrotic Mosaic Virus. *Comptes Rendus Acad. Sci.* Serie III, **299**, 53–8.

Mossé, J. and Pernollet, J.C. (1982) Storage proteins of legume seeds, in *Chemistry and Biochemistry of Legumes*, (ed S.K. Arora), Oxford and IBH, New Delhi.

Mukurumbira, L.M. (1985) Effects of rate of fertilizer nitrogen and previous grain legume crop on maize yields. *Zimbabwe Agric. J.*, **82**, 177–9.

Munene, F. (1982) A tough nut to grow. *IDRC Reports*, **11**, 19.

Muniyappa. V. and Veeresh, G.K. (1984) Plant virus diseases transmitted by whiteflies in Karnataka. *Proc. Indian Academy Sci: Animal Sciences*, **93**, 397–406.

National Research Council (1979) *Tropical Legumes, Resources for the Future*. National Academy of Sciences, Washington D.C.

Ndiokwere, C.L. (1982) Determination of crude protein and some mineral content of edible Nigerian legumes using neutron activation analytical technique. *Legume Research*, **5**, 87–90.

Ng, N.Q, Thottappilly, G. and Gumedzoe, M. (1985) Virus diseases. *Annual Report of the International Institute of Tropical Agriculture for 1984*, Ibadan, 223.

Nishitani, T. and Inouye, J. (1981) The growth habits of bambara groundnut

(*Voandzeia subterranea* Thouars) and Kersting's groundnut (*Kerstingiella geocarpa* Harms) – effects of day length and temperature on flowering. *Japanese J. Trop. Agric.*, **25**, 163–9.

Nishitani, T., Muraki, K. and Inouye, J. (1984) On the water storage cells in the leaves of bambara groundnut (*Vigna subterranea* (L.) Verdc.) and Kersting's groundnut (*Macrotyloma geocarpum* (Harms) Maréchal & Baudet). *Japanese J. Trop. Agric.*, **28**, 149–54.

Nishitani, T., Jin, I.D. and Inouye, J. (1988a) Developments of embryosac and embryo in bambarra groundnut (*Vigna subterranea* (L.) Verdc.) *Japanese J. Trop. Agric.*, **32**, 75–9.

Nishitani, T., Muraki, K. and Inouye, J. (1988b) Effects of daylength on the flowering and fruiting in bambarra groundnut (*Vigna subterranea* (L.) Verdc.) *Japanese J. Trop. Agric.*, **32**, 80–4.

Nnadi, L.A. (1980) Nitrogen economy in selected farming systems of the savanna region, in *Nitrogen Cycling in West African Ecosystems. Proceedings of a Workshop*, (ed. T. Rosswall), Royal Swedish Academy of Sciences, Stockholm, pp. 345–51.

Nnadi, L.A. and Balasubramanian, V. (1978) Root nitrogen content and transformation in selected grain legumes. *Trop. Agric.*, **55**, 23–32.

Nnadi, L.A., Balasubramanian, V. and Singh, L. (1976) Yield and mineral composition of grain legumes. *Ghana J. Agric. Sci.*, **9**, 211–15.

Nnadi, L.A., Singh, L. and Balasubramanian, V. (1981) Effect of grain legumes and sorghum on soil nitrogen status and the yield of subsequent maize crop. *Samaru J. Agric. Research*, **1**, 183–90.

Nwokolo, E. (1987) A nutritional assessment of African yam bean *Sphenostylis stenocarpa* (Hochst ex A. Rich) Harms. and bambara groundnut *Voandzeia subterranea* L. *Food and Agriculture*, **41**, 123–9.

Nyamudeza, P. (1989) *Crop Water Use and the Root Systems of Bambara Groundnut (*Vigna subterranea* (L.) Verdc.) and Groundnut (*Arachis hypogaea* (L.)) in Response to Irrigation and Drought*, M.Sc. Thesis, University of Nottingham.

Nyasaland Protectorate (1959) A summary of some aspects of agricultural experimental work, 1958/59. *Nyasaland Farmer and Forester*, **5**(1), 7.

Nyasaland Protectorate (1960) *Annual Report of the Department of Agriculture for the year 1958/59*, Part II, Government Printer, Zomba.

Nyasaland Protectorate (1961) *Annual Report of the Department of Agriculture for the year 1959/60*, (Part II), Government Printer, Zomba.

Nyasaland Protectorate (1963) *Annual Report of the Department of Agriculture for the year 1961/62*, (part II), Government Printer, Zomba.

Obizoba, I.C. (1983) Nutritive value of cowpeas – bambara groundnut – rice mixtures in adult rats. *Nutrition Reports International*, **27**, 709–12.

Obizoba, I.C. (1989) The nutritive value of mixtures of two varieties of bambara groundnut and sorghum. *Cereal Chemistry*, **66**, 249–51.

Ofuya, Th.I. and Bamigbola, K.A. (1991) Damage potential, growth and development of the seed beetle, *Callosobruchus maculatus* (Fabricius) (Coleopetera: Bruchidae), on some tropical legumes. *Trop. Agric.*, **68**, 33–6.

Ogbuji, R.O. (1979) Effects of two *Meloidogyne* spp. on growth and reproduction of bambara groundnut (*Voandzeia subterranea*) in Nigeria. *Tropenlandwirt*, **80**, 47–51.

Okigbo, B.N. (1973) Grain legumes in the farming systems of the humid lowland tropics, in *Proceedings of the First IITA Grain Legume Improvement Workshop*, International Institute of Tropical Agriculture, Ibadan, pp. 211–23.

Oliveira, J.S. (1976) Grain legumes of Mozambique. *Tropical Grain Legume Bull.*, **3**, 13–15.

Oluyemi, J.A., Fetuga, B.L. and Endeley, H.N.L. (1976) The metabolizable energy value of some feed ingredients for young chicks. *Poultry Science*, **55**, 611–18.

Ostendorf, F.W. (1962) *Nüttige Planten en Sierplanten in Suriname.* Landbouwproefstation in Suriname, Bulletin No. 79, 3rd edition.

Owusu-Domfeh, K., Christensen, D.A. and Owen, B.D. (1970) Nutritive value of some Ghanian feed-stuffs. *Canadian J. Animal Sci.*, **50**, 1–14.

Padulosi, S. and Ng, N.Q. (1989a) Collecting germplasm in Zimbabwe. *Plant Genetic Resources Newsletter*, **78/79**, 35–6.

Padulosi, S. and Ng, N.Q. (1989b) Collecting wild *Vigna* and *Oryza* in Tanzania. *Plant Genetic Resources Newsletter*, **78/79**, 39–40.

Padulosi, S., Lahgetti, G., Ng, N.Q. and Perrino, P. (1989) Collecting in Swaziland and Zimbabwe. *Plant Genetic Resources Newsletter*, **78/79**, 38.

Pereira, J. (1983) The effectiveness of six vegetable oils as protectants of cowpeas and bambara groundnuts against infestation by *Callosobruchus maculatus* (F.) (Coleopetera: Bruchidae). *Stored Products Research*, **19**, 57–62.

Phillips, R.D. and Abbey, B.W. (1989) Composition and flatulence-producing potential of commonly eaten Nigerian and American legumes. *Food Chemistry*, **33**, 271–80.

Poulter, N.H. (1981) Properties of some protein fractions from bambara groundnut [*Voandzeia subterranea* (L.) Thouars]. *J. Sci. of Food and Agric.*, **32**, 44–50.

Poulter, N.H. and Caygill, J.C. (1980) Vegetable milk processing and rehydration characteristics of bambara groundnut [*Voandzeia subterranea* (L.) Thouars]. *J. Food and Agriculture*, **31**, 1158–63.

Prevett, P.F. (1967) Notes on the biology, food plants and distribution of Nigerian bruchidae (Coleopetera), with particular reference to the northern region. *Bull. Entom. Soc. Nigeria*, **1**, 3–6.

Prevett, P.F. (1970) *Callosobruchus subinnotatus* (Pic) (Coleopetera, Bruchidae) – a potential pest of stored groundnuts. *J. Stored Products Research*, **6**, 279–80.

Rachie, K.O. and Roberts, L.M. (1974) Grain legumes in the lowland tropics. *Adv. Agron.*, **26**, 1–132.

Raheja, A.K. (1976) Insect pests of grain legumes other than cowpea grown at Samaru. *Samaru Agricultural Newsletter*, **18**, 98–101.

Rassel, A. (1960) Le voandzou *Voandzeia subterranea* Thouars et sa culture au Kwango. *Bulletin Agricole du Congo Belge et du Ruanda-Urundi*, **51**, 1–26.

Rowson, J.M. (1952) The hypodermal zone in the testa of certain leguminous seeds. *J. Royal Microscopical Soc.*, **72**, 46–55.

Ruthenberg, H. (1980) *Farming Systems in the Tropics.* 3rd ed, Clarendon Press, Oxford.

Salunkhe, D.K. and Kadam, S.S. (1989) *CRC Handbook of World Food Legumes: Nutritional Chemistry, Processing Technology, and Utilization*, Volume III, CRC Press, Boca Raton, Florida.

Seeger, J.R. and Maldague, M.E. (1960) Infestation de nodules de légumineuses en région équatoriale par des larves de *Rivellia* sp. (Dipt.). *Parasitica*, **16**, 75–84.

Sellschop. J.P.F. (1962) Cowpeas, *Vigna unguiculata* (L.) Walp. *Field Crop Abstracts*, **15**, 159–266.

Siemonsma, J.S. (1982) *La culture du gombo (Abelmoschus spp.), légume-fruit tropical avec référence spèciale a la Côte d'Ivoire.* Landbouwhogeschool, Wageningen.

Smartt, J. (1959) Some factors limiting the production of edible legumes in Northern Rhodesia, in *Proceedings of the 4th Annual Conference of Professional Officers*, Department of Research and Specialist Services, Government Printer, Salisbury, p. 13.

Smartt, J. (1985) Evolution of grain legumes. III Pulses in the genus *Vigna. Exper. Agric.*, **21**, 87–100.

Smyth, K.J.F. (1968) Bambara groundnut variety trials at Ukiriguru. *Ukiriguru Research Notes*, **21**, 1–2.

Somasegaran, P., Abaidoo, R.C. and Kumaga, F. (1990) Host-*Bradyrhizobium* relationships and nitrogen-fixation in the Bambarra groundnut [*Voandzeia subterranea* (L.) Thouars nom. cons.]. *Trop. Agric.*, **67**, 137–42.

Southgate, B.J. (1958) Systematic notes on species of *Callosobruchus* of economic importance. *Bull. Entom. Research*, **49**, 591–9.

Sreeramulu, N. (1982) Changes in growth regulator contents in developing seeds of bambarra groundnut (*Voandzeia subterranea* Thouars). *Plant Physiology and Biochemistry*, **9**, 113–18.

Sreeramulu, N. (1983) Germination and food reserves in bambarra groundnut seeds (*Voandzeia subterranea* Thouars) after different periods of storage. *J. Expl. Bot.*, **34**, 27–33.

Stanton, W.R., Doughty, J., Orraca-Tetteh, R. and Steele, W. (1966) *Grain Legumes in Africa*. Food and Agriculture Organization, Rome.

Stapf, O. (1912) Discovery of *Voandzeia subterranea* in the wild state. *Bull. Miscell. Information Royal Botanic Gardens Kew*, **26**, 213.

Steele, W.M., Allen, D.J. and Summerfield, R.J. (1985) Cowpea (*Vignea unguiculata* (L.) Walpi), in *Grain Legume Crops*, (eds R.J. Summerfield and E.H. Roberts), Collins, London, pp. 520–83.

Swaziland Government (1979) *Annual report of the Ministry of Agriculture and Cooperatives*, Mbabane.

Swaziland Government (1983) *Annual Statistical Bulletin for 1981*, Central Statistical Office, Mbabane.

Tardieu, M. (1958) Les cultures d'appoint dans la zone d'action du CRA de Bambey. *Bulletin Agronomique Ministère de la France d'Outre Mer*, **17**, 5–54.

Teyegaga, A. (1970) *Studies on the Development, Germination and Survival of Conidia of* Cercospora canescens *Ellis and Martin and the Development of Leaf Spot of* Voandzeia subterranea *Thouars, caused by* Cercospora canescens, M.Sc. Thesis, University of Ghana, Legon.

Thompson, E.J. and Dennis, E.A. (1977) Studies on nodulation and nitrogen fixation by selected legumes, in *Proceedings of the Joint University of Ghana – Council for Scientific and Industrial Research Symposium on Grain Legumes in Ghana*, (ed. E.V. Doku), University of Ghana, Legon, pp. 85–102.

Toms, G.C. and Western, A. (1971) Phytohaemagglutinins, in *Chemotaxonomy of the Leguminosae*, (eds J.B. Harborne, D. Boulter and B.L. Turner), Academic Press, London, pp. 367–462.

Umezurike, G.M. and Numfor, F.A. (1979) Changes in the content of starch and activities of some enzymes of carbohydrate metabolism in cotyledons of germinating seeds of *Voandzeia subterranea*. *J. Exper. Bot.*, **30**, 583–8.

Uwaegbute, H.O. (1977) The use of local feed ingredients in livestock feeds. *Canadian J. Animal Sci.*, **57**, 835.

Valentine, A.S. (1963) *Bambara Nut Advisory Notes*, Ministry of Agriculture, Southern Rhodesia.

Vaughan, J.G. (1970) *The Structure and Utilization of Oil Seeds*, Chapman and Hall, London.

Verdcourt, B. (1980) The correct name for the bambara groundnut. *Kew Bulletin*, **35**, 474.

Visser, J.H. (1975) Germination stimulants of *Alectra vogelii* Benth. seed. *Zeit. Pflanzenphysiol*, **74**, 464–9.

Visser, J.H., Dorr, I. and Kollmann, R. (1977) On the parasitism of *Alectra vogelii* Benth. (Scrophulariaceae) I. Early development of the primary haustorium and initiation of the stem. *Zeit. Pflanzenphysiol.*, **84**, 213–22.

Wallace, G.B. and Wallace, M.M. (1947) Second supplement to the revised list of plant diseases in Tanganiyika territory. *East African Agric. J.*, **13**, 61–4.

Westphal, E. (1981) L'agriculture autochtone au Cameroun: les techniques culturales, les séquences de culture, les plantes alimentaires et leur consommation. *Miscellaneous Papers Landbouwhogeschool*, no. 20, Wageningen.

Wolk, P.C. van der (1915) Een en ander over de aanaarding van katjang bogor (*Voandzeia subterranea*). *Cultura*, **27**, 405–17.

Wrigley, G. (1969) *Tropical Agriculture; The Development of Production*, Faber and Faber, London.

Zagorodsky, M. (1911) Die Erderbse (*Voandzeia subterranea* Thouars) und ihre Verwertung als Futtermittel. *Der Tropenpflanzer*, **15**, 413–36.

Zambia Government (1975) *Annual Report of the Research Branch 1973–74*. Department of Agriculture, Ministry of Rural Development.

Zulu, E.D. (1989) *Germination and Early Seedling Growth of Bambara Groundnut (*Vigna subterranea *(L.) Verdc.) in Response to Temperature, Sowing Depth and Soil Water*, M.Sc. Thesis, University of Nottingham.

2

Yam bean (*Pachyrhizus* species)

M. Sørensen, M. Grum, R. E. Paull, V. Vaillant, A. Venthou-Dumaine and C. Zinsou

2.1 HISTORY AND USES

Recent interest has focused on the neotropical genus *Pachyrhizus* Rich. ex DC. due to its known capability of producing high tuber yields and its considerable adaption to diverse climatic and edaphic conditions. The original spelling of the name by De Candolle with a single 'r' has been conserved over the earlier spelling with 'rr'. The generic name is from the Greek, *pachys* = thick and *rhiza* = root. The genus comprises five species of which three are cultivated for their tubers and two are exclusively wild. Of the cultivated species only the Mexican yam bean has been widely introduced pantropically but this cannot be explained by the lack of agriculturally attractive characters in the remaining two cultivated species.

Considerable research has been carried out since the NRC report (1979) stated that yam bean remained a primitive crop, with no concerted efforts to collect germplasm or use it, despite demands in many parts of the tropics. This account represents a synthesis of this research, which is continuing in several parts of the world.

The cultivated species include *P. erosus*, Mexican yam bean (Figure 2.1); *P. ahipa*, Ahipa (Figure 2.1); and *P. tuberosus*, Potato bean (Figure 2.2). Each appear to have been domesticated in separate areas: *erosus* in Mexico and north Central America, *ahipa* in Bolivia and N. Argentina, and *tuberosus* around the headwaters of the Amazon. *P. erosus* is the most important cultigen.

Pre-Columbian and early post-Columbian references to the cultivation history of the Mexican and Central American species, *P. erosus*, indicate that the 'Xicama' (Aztec), 'Mehen-chikam' (Maya) or 'Guyati' (Zapotec) was cultivated by all major civilizations including Toltec, Aztec and Mayan

Pulses and Vegetables. Edited by J. T. Williams. Published by Chapman & Hall, 2–6 Boundary Row, London SE1 8HN. ISBN 0 412 46610 4.

(a)

(b)

(c)

(d)

(e)

(f)

Figure 2.2 (a) a single *P. tuberosus* plant grown near Lamas, Prov. San Martin, Peru; (b) legumes of the *P. tuberosus* plant (same site as A); (c) local cultivar from Portoviejo, Prov. Manabi, Ecuador; (d) farmer with harvested legumes, Portoviejo.

Figure 2.1 (a) yam bean (*P. erosus*) harvest near Celaya, State of Guanajuato, Mexico, November 1988, yield approx. 70 tonnes ha $^{-1}$. Photo: M. Sørensen; (b) average size of yam bean (*P. erosus*) tuber for export to USA. Celaya, State of Guanajuato, Mexico. Photo: M. Sørensen; (c) yam bean harvest near Khon Kaen, Thailand; December 1988, yield approx. 20–30 tonnes ha $^{-1}$. Photo: O. Stølen. (d) *P. erosus* tubers from wild populations (left) near Xalapa, State of Vera Cruz, Mexico and cultivars from the State of Guanajuato, Mexico. Photo: B. Johansen; (e) *P. ahipa* originally from the Province of Tarija, Bolivia, grown in field trial at CIFAP/CEBAJ, Celaya, State of Guanajuato, Mexico. Photo: M. Grum; (f) sliced tubers of *P. ahipa*. This species has a unique purple colouring. CIFAP/CEBAJ, Mexico. Photo: M. Grum.

according to Urbina (1906), Standley (1930) and Martinez (1979). The relative uniformity seen in cultivars in regions outside the distribution area of the species, e.g. the Yucatan peninsula (Standley, 1930) and the southern departments of Guatemala indicates that these cultivars are remnants of earlier cultivars introduced from the central or southern parts of Mexico. In the Guatemalan department of Peten, *P. erosus* is still to be found as one of the crops grown in shifting cultivation and this is the northernmost locality of this cultivation practice.

The present day Mexican cultivars, primarily used in the state of Guanajuato and Nayarit where the crop is either intercropped with maize (*Zea mays* L.) and common bean (*Phaseolus vulgaris* L.) or monocropped, are morphologically very uniform when compared to the local 'landraces' from, e.g., Oxaca and Vera Cruz (Heredia Zepeta, 1985). Following plant movements on Spanish galleons to the Philippines and later distribution by the Royal Botanic Gardens, Kew, to tropical British colonies, the Mexican yam bean (*P. erosus*) is now known in cultivation in large regions outside its original distribution area, e.g. in Southeast Asia, India, the Pacific and the West Indies (Sørensen, 1990). The history of the spread of the crop is recorded: for instance it was introduced into Indonesia from Manila through Aubon in the Mollucas at the end of the 17th century. *P. erosus* tubers are found in vegetable markets in the Philippines, Indonesia, Malaysia, Thailand, Myanmer, Taiwan and in China. In China, in the provinces of Sichuan and Chengdu, *P. erosus* or soyabean are usually planted on the ridges between rice and paddy fields. This practice is unknown in Thailand where these ridges are usually kept cleared as a precaution against rats.

Ahipa is much more localized in distribution, being limited to Andean valleys in Bolivia and possibly Peru where the crop is virtually extinct today. Potato bean is distributed from the Southern lowlands of Colombia and Venezuela where it is known as 'nupe', through the Amazonian parts of Ecuador ('jiquima'), Peru ('jiquima', 'ashipa' or 'dabau'), Bolivia ('ajipa' or 'cooco') and Brazil including the northeastern lowlands of Paraguay (in Brazil and Paraquay the local name is 'mbacucu').

P. tuberosus is the only other species which has been introduced to areas outside the continent in pre-Columbian times. The species is recorded from several of the greater Antilles.

The crop produces a fleshy underground tuberous root with a thick brownish skin. The flesh is white and crisp and has a sweet flavour. The flesh is either used fresh as slices in both green and fruit salads or it is cooked so that the crisp texture remains. In some areas the tubers are used in pickles, e.g. in Java. Cooked tubers are handled in a similar way to potatoes. Tubers can reach up to 3 kg in weight but contain less dry weight than other tuber crops. Apart from the tubers, the rest of the plant is not valuable for food, but occasionally the young tender pods of *P. erosus* are boiled and eaten as a vegetable. The young pods of *P. tuberosus* are not eaten because of the presence of irritant hairs.

Speciality markets exist for yam bean. Mexico exports to the USA to supplement scarce water chestnuts used in Chinese cooking and for low-calorie snacks (NRC, 1979).

Starch can be extracted from tubers and is used in confections, resembling as it does arrowroot starch (Porterfield, 1939). The crop has also received attention as a green manure crop in Southeast Asia (Masefield, 1957).

2.2 TAXONOMY AND ECOLOGY

The genus is placed in the subtribe Diocleinae of the tribe Phaseoleae of the Leguminosae. Although the genus had been the subject of revision by Clausen (1945), the taxonomy remained, until recently, somewhat diffuse, mainly due to the limitation of available herbarium material caused by World War II. Clausen applied a narrow species concept and included a great number of infraspecific taxa. Better availability of the herbarium material available from all relevant European herbaria as well as materials collected during the past 40 years led to a second revision (Sørensen, 1988).

The genus is morphologically delimited by the presence of the following characteristics: Annual or perennial vines or semi-erect herbaceous to somewhat lignified plants with one or more tuberous roots; trifoliate leaves with stipules, pinnately arranged leaflets with caducous stipels; the inflorescence a complex to simple raceme; flowers with calyx and a papilionaceous corolla; the ovary has a basal crenulate disciform nectarium; the recurving style is ciliated ventrally and the vertical surface of the stigma is subglobose; the straight legume is septate between the seeds and the 4–9 seeds are square, more or less flattened or rounded kidney-shaped with colours olive green, deep maroon, red, purple, black or black and white mottled.

Chemosystematic analysis of the phylogeny and interrelationships at the generic as well as the subtribal level have so far been limited to studies of canavanine (Lackey, 1977) and isoflavonoid phytoalexins (Ingham, 1979). These studies have been broadened by new research on isozymes, chloroplast and ribosomal DNA at the Department of Preclinical Medicine and Biology, University of St. Andrews, Scotland, UK and the Botanical Section, Department of Botany, Dendrology and Forest Genetics, Royal Veterinary and Agricultural University, Copenhagen, Denmark. Palynological studies (Sørensen, 1989a) revealed morphologies permitting unequivocal species identification, with large intraspecific variation in the pollen grains of the cultivated species.

Yam beans grow well in hot wet areas although they grow in locations ranging from subtropical to tropical and dry to wet. They tolerate some drought but are sensitive to frost (NRC, 1979). Light, well-drained soils are preferable.

P. tuberosus is also known wild from Ecuador and Peru and of the three cultivated species it tolerates higher precipitation than the others, where under such conditions, cracking or splitting of the tuber can be a problem. *P. ahipa* has no wild material and shows considerable adaptability to climatic variation and trials have been carried out in Tonga and under flood irrigation in Mexico.

There are two related wild species, *P. ferrugineus*, associated with evergreen to deciduous rainforest along the Atlantic coast of Central America from Vera Cruz, Mexico to the Colombian department of Choco on the Pacific coast. Soils are usually low in available phosphorus, unusual for a tuberous plant (Sørensen, 1990). No hybridization has been attempted between this species and the cultigens. The second wild species, *P. panamensis* is found in seasonally dry regions of central and Pacific Panama projecting into the coastal deciduous forest of the Department of Santa Marta in Colombia, and also in central Ecuador in Provinces Guayas and El Oro. This latter species exhibits adaption to seasonal drought and successful hybrids have been made with *P. tuberosus* (reciprocally) and with an interspecific *ahipa* × *tuberosus* hybrid.

The altitudinal range of *P. erosus* is shown in Figure 2.3 which shows that the natural variation accessed into a germplasm collection accords with one measure of ecogeographic differentiation as recorded on botanical specimens.

2.3 MORPHOLOGY AND CYTOLOGY

2.3.1 *Pachyrhizus ahipa* (Wedd.) Parodi

Ahipa or Ajipa is probably the most interesting of all the species from the systematic point of view because of the absence of known wild material, although Brücher (1989) suggests that the type known today was probably selected from wild material growing in Ceja de Montana in Peru. It is also interesting because morphologically it is erect with short growth habit and because of daylength neutrality has a short growth season (five months).

P. ahipa is non-climbing and furthermore is distinguished by its entire leaflets, short racemes (48–92 mm) and reduced number or absence of lateral racemes. The number of flowers per lateral raceme is as low as 2–6. The outwards curling wings are likewise a feature seen only in *P. ahipa*. Seeds are black or black and white mottled, reniform and 9 × 10 mm.

2.3.2 *Pachyrhizus tuberosus* (Lam.) Spreng.

P. tuberosus is recognized by the following morphological characters: it is a climber and is the largest species with stems to 5–7 m and terminal leaflets to 280 × 260 mm. The legumes are also larger than those of the other species, to 255 × 23 mm, conspicuously compressed between seeds. The

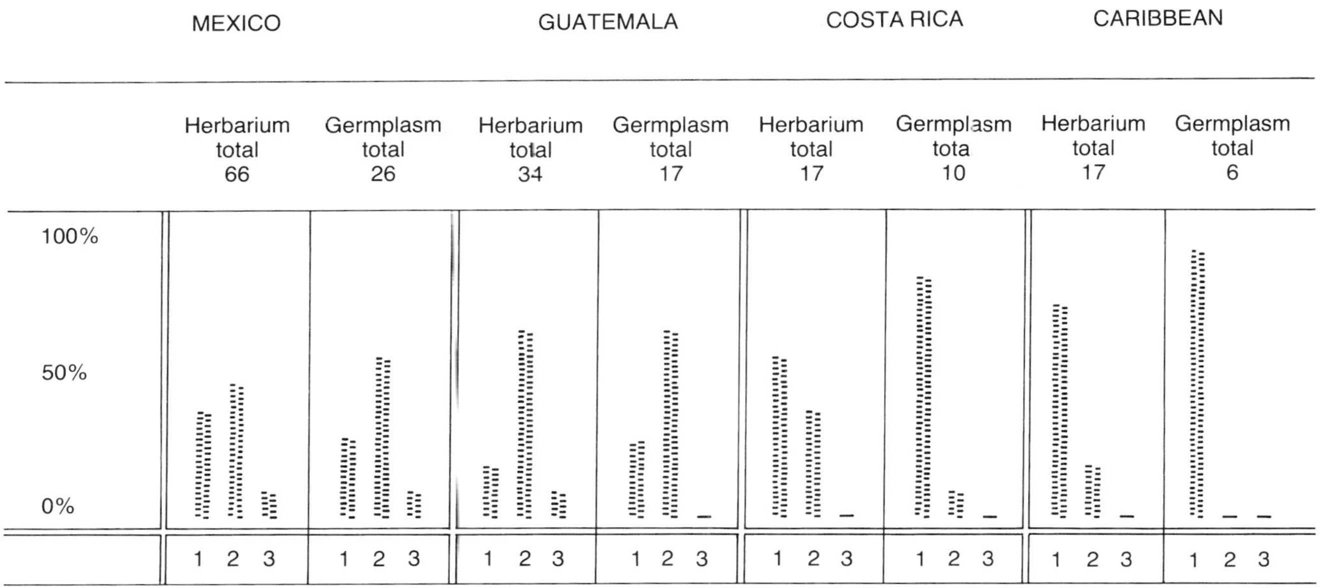

Figure 2.3 The altitudinal range of herbarium material versus that of germplasm material of collected *P. erosus*. **1**, 0–500 m, **2**, 500–1500 m, **3**, 1500+ m above sea level. Total neotropical herbarium specimens seen: *P. erosus*, 249; *P. ferrugineus*, 124; *P. panamensis*, 19; *P. tuberosus*, 61 and *P. ahipa*, 8. Total accessions in germplasm collection: *P. erosus*, 85 (63 cult. and 22 wild); *P. ferrugineus*, 13; *P. panamensis*, 1; *P. tuberosus*, 11 and *P. ahipa*, 4.

seeds are black, black and white mottled or orange-red and reniform up to 12 × 14 mm.

Wild material studied has been limited. Cultivars appear to fall into two distinct groups: (1) those with long internodes, entire leaflets and long racemes (up to 360 mm) and (2) those with short internodes, deeply-lobed leaflets and racemes interrupted by leaflets with abrupt alternately left and right turning growth. The first group is known from the entire distribution in the Americas and the second group is peculiar to Ecuador. Daylength neutrality has been recorded in a number of cultivars.

2.3.3 ***Pachyrhizus erosus* (L.) Urban**

Morphologically *P. erosus*, also a climber, is identified by the lack of hairs on the petals, the number of flowers per lateral raceme (4–11), and the legume characters: colour pale olive to light brown and size 60–130 mm × 8–17 mm. The seed shape, square to rounded flattened but never reniform, also serves to identify the species, (Sørensen, 1988 and 1990).

2.3.4 ***Pachyrhizus ferrugineus* (Piper) Sørensen**

This wild species is the only one in the genus which is evergreen and the above-ground parts of the plant are perennial. Both tuber and stem are more woody than in the other species. The root is less tuberous (although plants grown in a glasshouse have produced fair sized tubers, ± 0.5kg). The outline of the leaflets are as variable as in *P. erosus*. The morphology of the inflorescence differs from the other species in the high number of flowers per lateral raceme: 8–21; the length of the main axis of raceme is up to 860 mm. The length/width ratio of legume is distinctly different: length 80–130 mm and width 12–23 mm; also the prominent reddish-brown strigose hairs and the lack of constrictions between the seeds are distinguishing characters. The seeds are rounded 13 × 13 mm and laterally compressed.

2.3.5 ***Pachyrhizus panamensis* Clausen**

Morphologically the species is recognized by the white to light brown pilose hairs covering all parts of the plant, and the low number of flowers per lateral raceme: 4–7. The seeds are the smallest in the genus, rounded to slightly reniform, 6 × 7 mm. *P. panamensis* has produced fair sized tubers in the glasshouse as well as in field trials in Benin (± 200g). In 1991 field trials have been initiated in Senegal in order to test the drought tolerance of this species. Together with wild material of *P. erosus* from the province of Guanacaste in Costa Rica, *P. panamensis* probably constitutes the best genome adapted to seasonal drought.

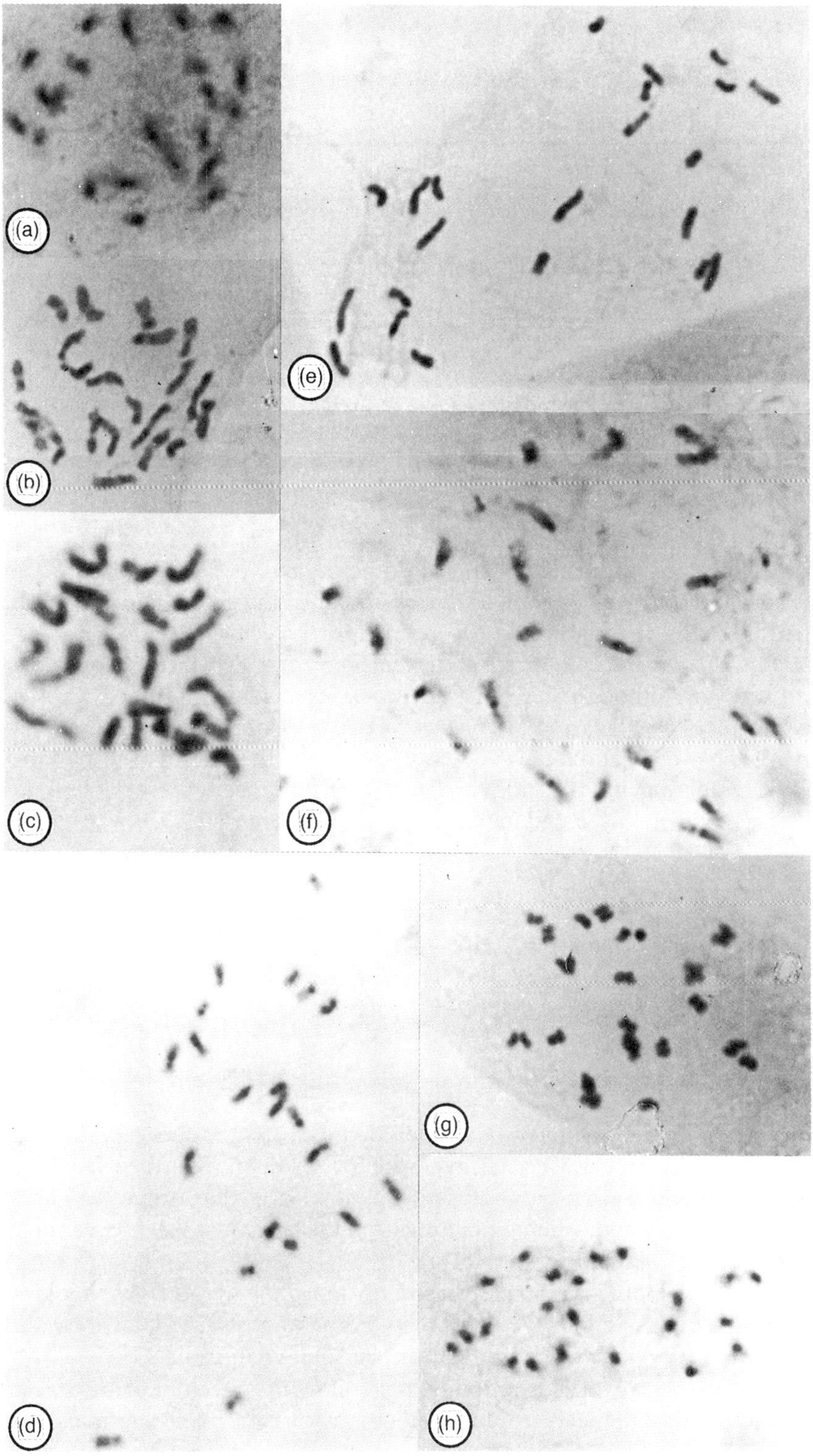

Figure 2.4 Cytological observations of mitotic cells from root tips of species.
(a) *P. erosus*, cultivar acc. no. EC107/1086; (b) *P. erosus*, wild form acc. no. EW048/385; (c) *P. erosus*, wild form acc. no. EW115/1285; (d) *P. ahipa*, cultivar acc. no. AC102/485; (e) *P. panamensis*, acc. no. PW055/385, (f) *P. ferrugineus*, acc. no. FW044/385; (g) *P. tuberosus*, cultivar acc. no. TC118/186; (h) *P. tuberosus*, cultivar acc. no. TC526/287. All illustrations × 2800.

Table 2.1 Developmental stages of *Pachyrhizus*

Reference	*Stage*	*Description*
V0	Germination	Appearance of the primary root
V1	Emergence	Primary leaves and epicotyl emerge
V2	Primary leaves totally opened	Primary leaves fully unfolded
V3	First trifoliate leaf opens	2nd trifoliate leaf appears
V4	Third trifoliate leaf opens	Buds at lower nodes produce branches
R5	Preflowering	First bud or first raceme appears
R6	Flowering	First flower opens
R7	Pod formation and filling	At the end of this stage the pods feel hard
R8	Post-pod growth	Pods filled and are hard
R9	Physiological maturity	Leaves senesce and defoliation starts

All species studied show chromosome number of 2n = 22 (Figure 2.4). This agrees with all reports of the x number of the genus (Lackey, 1981). Additionally a programme to produce interspecific hybrids among *P. erosus*, *P. ahipa* and *P. tuberosus* show that in general there is no major barrier to crossing although this is based on experimental conditions. Crosses between *P. tuberosus* and *P. ahipa*, however, were not successful. There is a barrier between cultivated and wild *P. erosus* and there seem to be intraspecific rather than interspecific barriers. The importance of these possibilities in breeding are discussed in Section 2.7 and in Sørensen (1989b).

2.4 PHYSIOLOGY OF GROWTH AND DEVELOPMENT

Most of the information available relates to *P. erosus*, the most widespread cultigen. In general, it is a high yielding crop with high yield stability and having a highly efficient symbiosis with nitrogen-fixing bacteria.

2.4.1 **Developmental patterns up to flowering**

The development patterns have been deduced from recent experimental plantings, in 1984 and 1989 (Zinsou 1988, Zinsou and Venthou-Dumaine 1988) and the experiments of Paull *et al.* (1988); the former based on February, May and October plantings and the latter two-weekly plantings from April to December.

The developmental stages, based on morphology and physiological changes, developed for *Phaseolus* bean (Fernandez *et al.*, 1986) were modified to fit *Pachyrhizus* (Grum, 1990) and are shown in Table 2.1.

The growth proceeds through the following developmental stages: Under conditions met in tropical regions whatever the photothermoperiod, tuberization (tuber initiation) occurs 4–6 weeks after germination (which itself takes 5–12 days) and root enlargement may be visible at the end of the 8th week (Table 2.1). Under conditions suitable for nodulation the first nodules appear after *ca.* 4 weeks.

Development under short days

When yam bean completes its cycle under short days (daylength < 12 h) flower initiation is early and coincides with tuber initiation. The V4 stage is very short or missing. At the end of 8 weeks the first flower buds (stage R5) are seen and at the same time root enlargement begins. The overlap of flowering and tuberization periods under short days was confirmed in *Pachyrhizus erosus* by Paull *et al.* (1988). As a consequence, shoot growth, flowering and tuber growth compete. In this case a twining vine-like growth habit is not induced. The size of the plant is limited and similar to that of dwarf bean. The plant allocates high priority to tuber production. When observed under short days *P. erosus* varieties, although they have an indeterminate shoot system, will develop plant sizes resembling plants with determinate growth due to the drastic reduction in the duration of the V4 stage or its complete lack.

Development under long days

When the growth cycle begins under long days the whole cycle covers successively long and short days and can last for 9 months. In this case tuber initiation occurs at 4–6 weeks after germination as under short day conditions. The long days lengthen the duration of the V4 stage causing shoot growth with formation of many twining vine-like shoots and root growth with a poor bulking. Flower initiation occurs when the daylength decreases and approaches 12.5 h.

Development under short and long days

When the cycle starts under short days (daylength < 12 h) but is not completed until the daylength gets longer than 12 h. the short day behaviour may be interrupted at any stage and replaced by long day behaviour. For instance in a plant sown in February (short day in the Northern hemisphere) both flowering and tuberization are initiated. Flower anthesis occurs but pod setting will be very poor with the increasing daylength. Tuber enlargement that began during short days stops. The plant reverts to the vegetative stage (from R5 stage to V4 stage). Shoot growth and root growth resume until the next short day period. The cycle will carry on as described above until the time of the second flower initiation.

Development after flowering

In all cases, even if the tuber bulking is reduced or stopped during shoot growth (effect of long days), it resumes and proceeds until the beginning of pod setting. Then pod setting and filling (stage R7) compete with tuber bulking until seed formation has been completed. Tuber bulking is completed when the legumes have filled and have become hard (end of stage R7). Physiological maturity of the plant (stage R9) described as the stage when leaf senescence and defoliation commence, does not intervene immediately with tuber bulking after legume maturity, but is delayed. It often occurs several weeks later at the end of tuber bulking in contrast to the development found in grain legumes.

Flowering, the fructification process and tuber bulking are closely controlled by short days (daylength < 12 h). Photoperiods of 14 to 16 h occurring in temperate regions prevent formation of the tuberous root in *P. erosus* (Kay, 1973). The shoot development (observed at the V4 stage) which increases the formation of photosynthetic metabolites is the only process where duration and quantity can be modulated and emphasized during long days.

Hence, as shoot development determines seed or tuber yield it is important that the plant is exposed to a sufficiently long period exceeding 12 h of daylight before flower induction reorientates the assimilate distribution to the storage sinks. In *P. tuberosus*, the influence of temperature and photoperiod has been described by Alvarenga and Valio (1989). They reported that only intermediate daylength (within 9 and 16 h) induced flowering. Tuberization took place only under daylengths of 16 h. Furthermore, they reported that day/night temperature of 30/25° delayed flower induction and inhibited tuberization. The completion of the different processes (flowering, pod setting and filling, tuber enlargement) were not examined and more work is needed in order to compare the two species.

Effects of growth factors

The effects of two growth regulators: gibberellic acid (GA) and chloride of chlormequate (CCC) sprayed on field plant were investigated through the developmental stages during both long and short day conditions.

Under short days, test plants were sprayed with both growth regulators after floral and tuber induction (6 weeks). GA (50 p.p.m. and 200 p.p.m.) enhanced shoot formation (long day effect) as well as tuber enlargement (short day effect), but applications severely reduced flowering and pod setting (long day effect). CCC (1000 and 5000) increased flowering and pod setting and also the number of seeds produced per plant, but reduced tuber bulking (long day effect). In each case the growth regulator diminishes the influence of the one sink for the benefit of the second one.

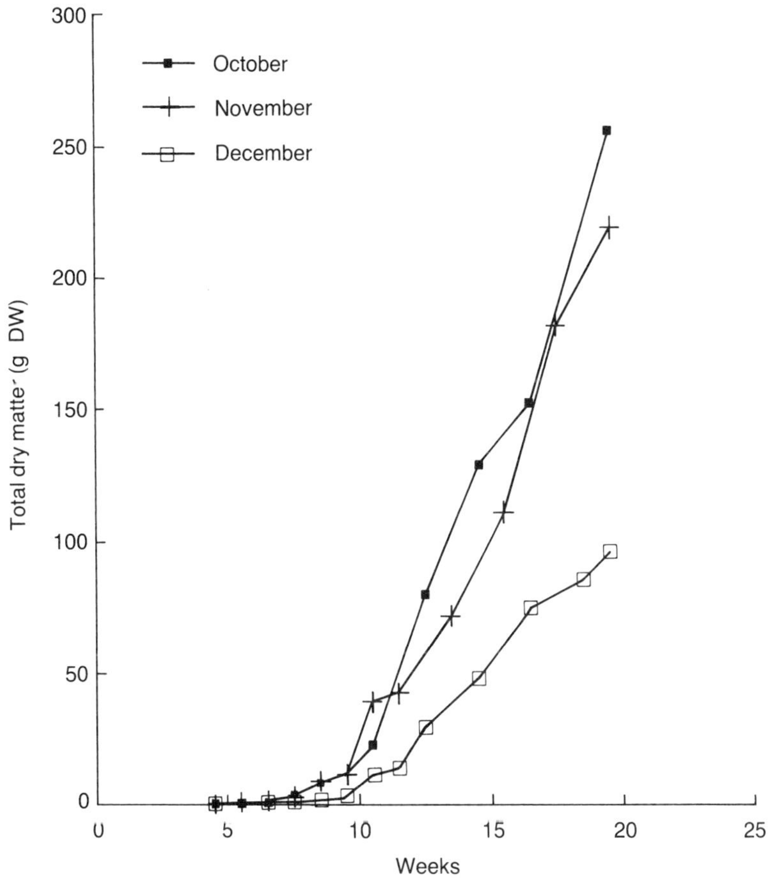

Figure 2.5 Accumulation of total dry matter per plant in plants grown in the field during short days (October, November and December plantings). The plants are staked to facilitate sampling. Each value is calculated from a minimum of ten plants.

Under long days the effects of the regulators appear to be more diffuse and thus more difficult to interpret.

2.4.2 Growth of the tubers

As growth and development of yam bean are sensitive to daylength, periods for tuber production have been determined for each region according to precipitation regimes and photoperiod.

Generally short days and moderate precipitation rates are the optimal conditions for tuber production; however, yam bean (*P. erosus*) can be grown successfully in the humid tropics (van Hoof and Sørensen, 1989).

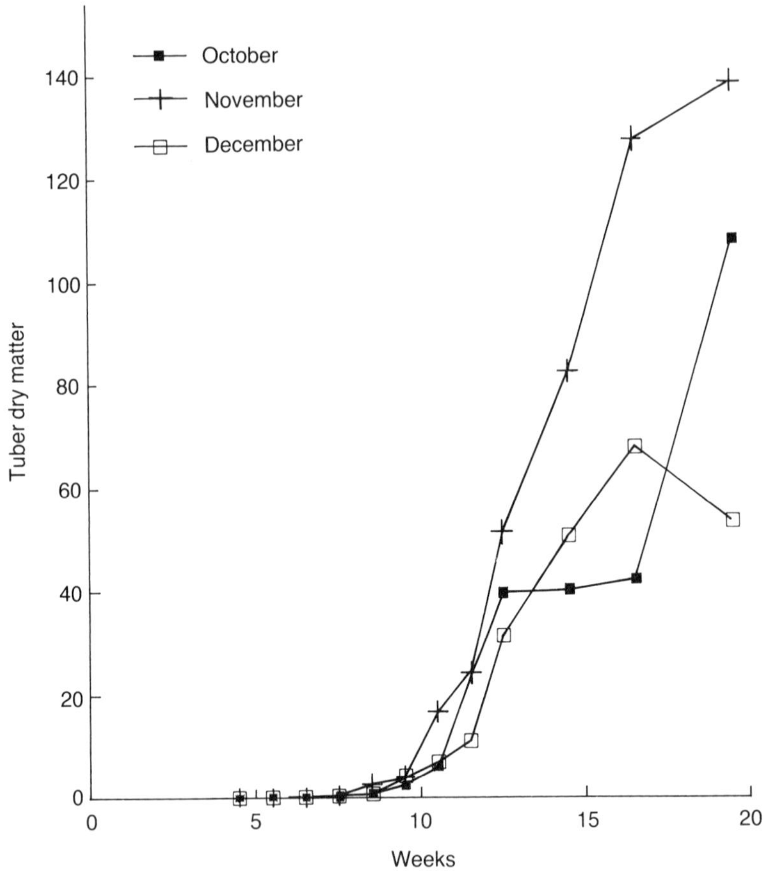

Figure 2.6 Accumulation of tuber dry matter per plant in plants grown in the field during short days (October, November and December plantings). The plants are staked to facilitate sampling. Each value is calculated from a minimum of ten plants.

Three types of growth sequences according to the three developmental patterns can be expected. Experimental results are summarized below.

Effects of short days

Plants from October, November and December plantings completed their cycles during short days. Cycle duration does not exceed five months. Because tuber growth and flowering rapidly compete with shoot development plants produced a photosynthetic apparatus of limited size and hence a poor dry matter accumulation in the whole plant. Shoots developed 30 ± 10 leaves per plant. Paull *et al.* (1988) reported the same number after 6 months for plants which had been induced to form tuberous roots at an

Table 2.2 Distribution, as percentage of total dry weight, at the end of 20 weeks, of dry matter allocated to the three compartments: pods, leaves plus stems and tuberous root of plants of *Pachyrhizus erosus* grown under short days (October planting) with various treatments

		Treatments (p.p.m.)			
Treatments	*Control*	*CCC 1000*	*CCC 5000*	*GA 50*	*GA 200*
Total dry weight /plant (g)	256	312	251	470	280
Pods DW /plant as % TDW	37	52	61	15	0.3
Tuberous root DW /plant as % TDW.	42	27	16	64	71.7
Leaves + stems DW /plant as % TDW	21	21	23	21	28

CCC, chlormequate; GA, gibberellic acid.

early stage. The development of the dry matter accumulated in the whole plant and in the tuber alone is shown in Figures 2.5 and Figure 2.6 in plantings from October, November and December.

In the graph representing the accumulation of dry matter in the tubers of October plants, competition between pod filling and tuber bulking reaches a plateau. Tuber bulking resumes after pod filling until shoot senescence. The competition between pod filling and tuber bulking appears to be non-existent in plants from November and December which could possibly be explained by a poor pod formation. GA (50 p.p.m. and 200) increased dry matter accumulation in the tuber while CCC (1000 and 5000 p.p.m.) did not. There were no significant differences in the dry matter accumulated in the whole plants between CCC (1000 and 5000) and GA 200 treatments. GA 50 increased the dry matter content of the whole plant (Table 2.2).

The effects of growth substances resulted in a different distribution of dry matter between pods and tubers. When the two sinks compete, flower abscission and reduced pod setting is caused by GA, i.e. tuber bulking is favoured. Inversely CCC enhanced pod setting and filling, but depressed tuber bulking. Under short day conditions CCC was used to improve seed yield.

Figure 2.7 shows the distribution ratio of the dry matter between the above and below ground parts of the plants. The distribution of the assimilates is preferentially orientated towards the shoot growth until the appearance of the first flower bud. Then the tuber becomes the major storage sink until pod setting. A second peak appears between 15 and 18 weeks in both control and CCC treated plants. The ascending part of the graph represents the pod filling stage and the descending one, the R8 stage. The

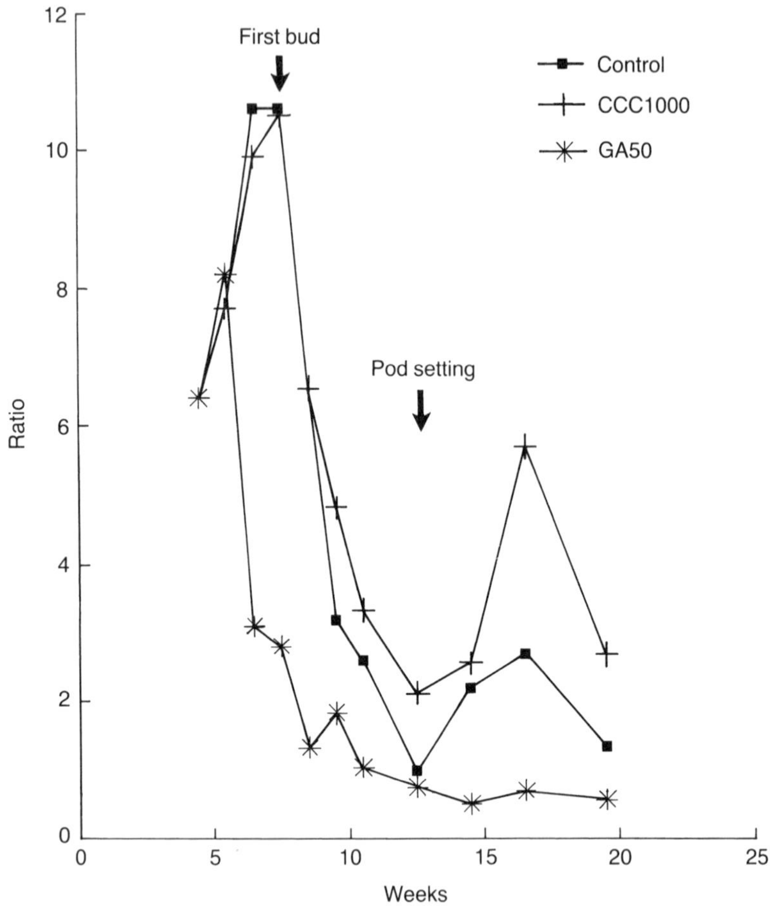

Figure 2.7 Development in dry matter distribution between shoot and underground parts (ratio) of plants treated by chloride of chlormequate (CCC, 1000 p.p.m.) and gibberellic acid (GA, 50 p.p.m.)

tuber bulking is resumed after pod filling. The absence of the peak in the graph representing the GA treated plants is due to the suppression of the pod sink by that growth factor.

Effects of long days

Plants from May took advantage of long days which inhibit flowering and tuber enlargement to develop shoots with high leaf number and large leaf area and roots that resulted in a large source, i.e. photosynthetic apparatus. The mean leaf area and leaf number per plant were 6.7 (± 0.3) and 165 (± 30) respectively after 18 weeks of growth. After 20 weeks it proved very difficult to continue the numbering of the leaves. The highest yield and tuber size is to be expected under conditions when the plants are sown and grown during long days and are cultivated on trellises with repeated removal of flowering shoots in order to prevent competition between pod filling and tuber bulking.

Effects of short and long days

Plants from February planting showed a double behaviour: A poor growth resulting from the effects of short days until April followed by an increased growth induced by long days from April onwards.

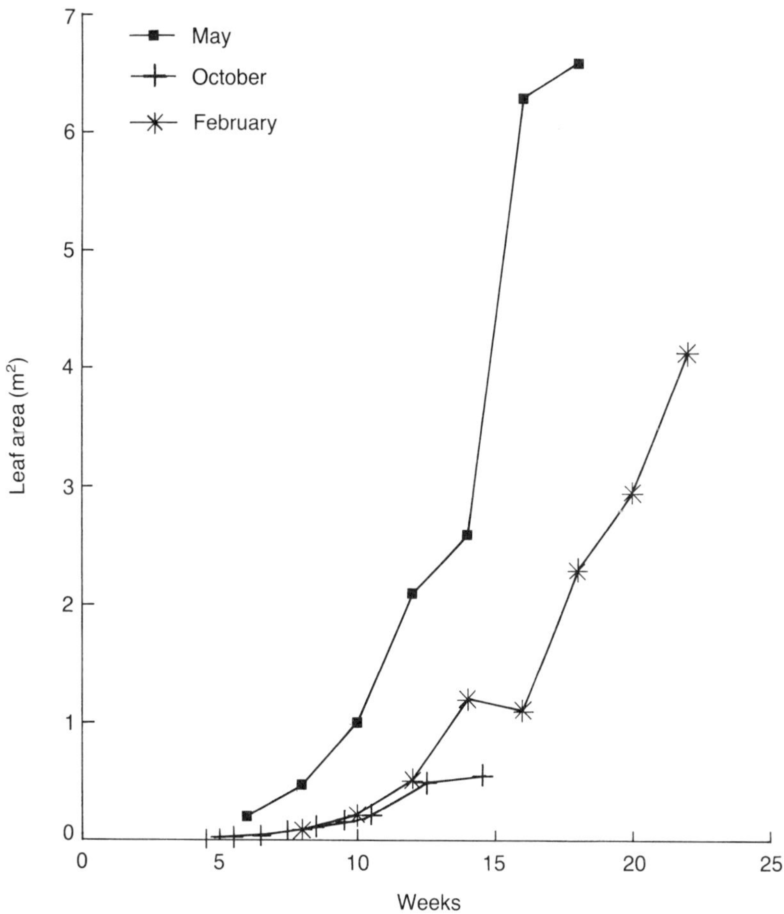

Figure 2.8 Comparative development of leaf area in plants from three different planting dates: May, October and February. Each value is the mean of ten plants, expressed in square meters.

Figures 2.8 and 2.9 show the comparative evolution of leaf area and dry matter accumulated per plant in October, May and February plantings. Comparisons of the dry matter accumulated during the first 18 weeks in the three types of growth, show that they are not significantly different.

Differences become obvious in the period between the 10th and 18th week. An early floral induction stimulated an early production of assimilates directed to storage sinks (pod and tuber) under short days. Under long days the assimilate production is progressive and it is allocated to develop the photosynthetic apparatus (source) and the tuber (sink) during the V4 stage. The longer the period of long days, the greater the shoot development which will later form the basis for the size of the pod sink and the assimilates devoted to pod filling and tuber bulking.

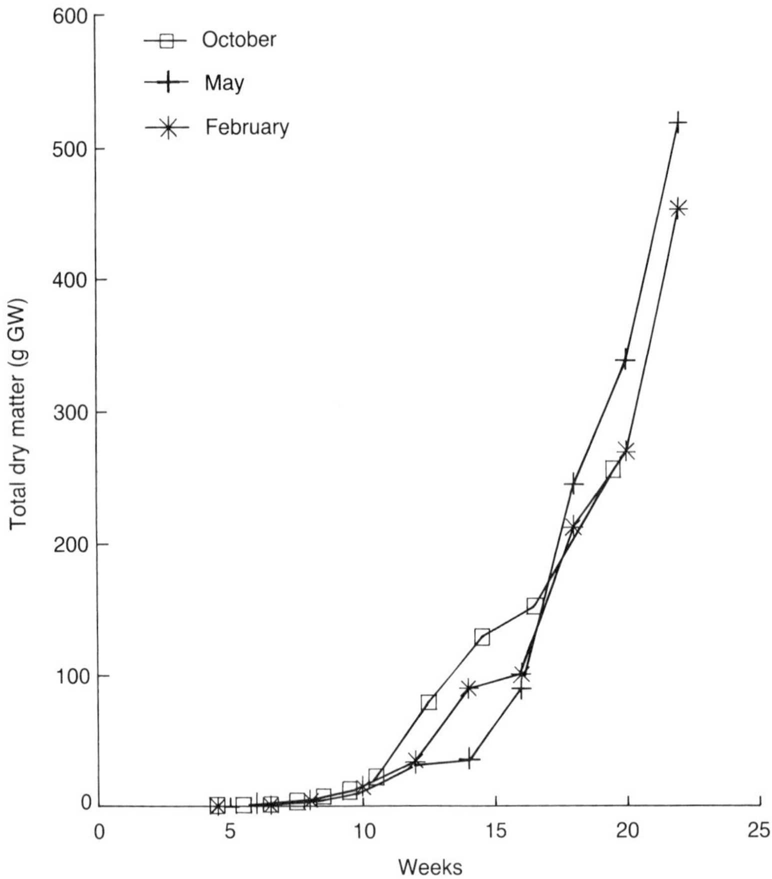

Figure 2.9 Comparative development of total dry matter accumulated in plants from three different planting dates: May, October and February.

2.4.3 **Source-sink relationships**

Laboratory investigations to better understand the role of sources and sinks in controlling the partitioning of recent photosynthates within the plant have been conducted using a ^{14}C pulse technique at the V4 stage (12 weeks) under long days when tuber bulking was inhibited.

Robin *et al.* (1990) found that apex and roots were the main sinks for carbon. Sixteen per cent of the newly assimilated carbon was temporarily stored as leaf starch. The kinetics of the distribution of ^{14}C showed that, after 24h, 27% and 38% of apex radioactivity was encountered in starch and protein respectively. The use of recent photosynthetic assimilates for maintenance and growth of nodules was low, whereas the specific activity of sugars decreased rapidly in nodules. Thus carbon allocation to nodules is primarily used to sustain the assimilation of fixed nitrogen.

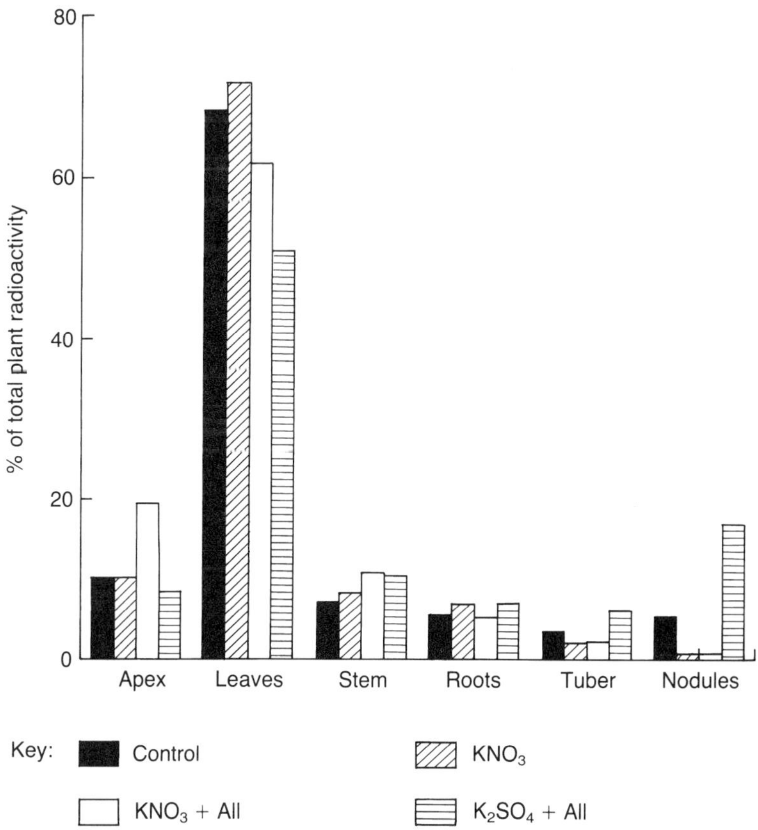

Figure 2.10 Radioactivity of various parts of yam bean (*P. erosus*) plants. Each value is the mean of three replicates (i.e. different plants). Data are expressed as the percentage of the total plant ^{14}C.

Vaillant *et al.* (1990) showed that shoot apices and nodules were still major sinks at the V4 (12 weeks) at the end of long days (September). By comparing nitrate-fed plants with control plants they found that nitrate depressed nodule and tuber growth. The analysis of the pattern of ^{14}C distribution showed that the large decline in nodule total radioactivity caused by nitrate resulted in only 36% decline in specific radioactivity of nodules (Figure 2.10).

Although tuber growth per plant was reduced by 50%, tuber specific activity was unaffected. The negative effect of nitrate on nodule and tuber growth was in agreement with the carbohydrate-deprivation hypothesis as a main cause for the inhibition of tuberous legume nodules by nitrate.

2.4.4 **Nitrogen nutrition and symbiosis with *Rhizobia***

Yam beans engage in symbiosis with different strains of *Rhizobia* of the cowpea group to fix atmospheric nitrogen. Woomer (1979) tested 23 strains of *Rhizobium* and found two with high efficiency, which gave a total nitrogen yield higher than a treatment with nitrate. After isolation of 15 strains from nodules harvested at different stages of development of the plants and different types of soils from Guadeloupe, Vansuyt and Zinsou (1991) studied the interactions among *Rhizobium* strains and soils. In the acid soils the *Rhizobia* are present. The absence of nodules or the expression of the symbiosis is only due to pH (4.6). The combined nitrogen is transported or stored as ureides, allantoin and allantoic acid (Pate *et al.* 1980). That trait is considered by some authors as a property of the tribe Phaseoleae (Zengbé and Salsac 1983).

As *Phachyrhizus* belongs to Phaseoleae, Lamaze *et al.* (1985) investigated the distribution of nitrogen compounds and purine catabolism in *P. erosus*. Their findings confirmed that the ureides represented the main forms of combined nitrogen transported. The ureide content peaked twice, the first peak at the beginning of tuber enlargement, and the second one at the pod setting. The enzymes involved in the degradation of the purines, responsible for formation of the ureides, likewise had two peaks of activity, at the beginning of tuberization and at pod setting. The allantoicase activity, essentially encountered in the stems, seems to confirm the hypothesis that ureides are converted into asparagine, aspartate, and glutamine during their transport from nodules to leaves and the aerial sinks (pods and seeds).

Zinsou and Vansuyt (1991) investigated the consequences of purine degradation inhibition induced by allopurinol on growth, nitrogenous compounds and carbohydrates in *P. erosus*. The plants grown in pots were treated by soil drench of allopurinol (0.4 mM during 14 consecutive days). They exhibited a decrease in xanthine and uricase activities in the nodules. As a consequence growth is reduced, the ureides level dropped in the nodules, tubers, stems and leaves, whilst sucrose and starch accumulated in the nodules. The decrease in the formation of ureides caused a decrease in the supply of nitrogen available for plant growth. The accumulation of sugars in the nodules can be explained by a fall in the sugar metabolism in the nodules.

This interpretation is confirmed by ^{14}C labelling studies and analysis of photosynthetic partitioning and utilization in nodules (Vaillant *et al.* 1991). In contrast to nitrate, allopurinol treatment appeared to have few effects on photosynthate allocation to the nodules. Due to the inhibition of ureide synthesis, radioactivity increased in nodules in relation to sucrose and starch accumulation. Lynd and Purcino (1987) investigated the nodulation and nitrogen fixation of *P. erosus* in relation to soil fertility. Total ureide content of tuber epiperidermal tissues increased significantly with P, Ca and P+Ca treatments.

2.4.5 **Conclusions**

Both the tuberization process (tuber induction to complete tuber bulking) and the flowering process (floral induction through pod filling) are severely influenced by short days. Only shoot growth may be adjusted, modulated or increased by the length and the duration of long days. High yields of tuber or seeds are determined by the size of the photosynthetic source (leaves and stems), and a daylength longer than 12 h for a certain duration prior to floral induction is needed. These physiological behaviours justify the dates of plantings recommended: March for the biggest tubers and June for marketable ones, in Mexico (Heredia Zepeda, 1985); September/October in Hawaii (Paull *et al.*, 1988).

Growth factors (GA and CCC) may be used to modify the size of the source and the direction and the quantity of assimilates allocated to the different storage sinks. GA spraying favours tuber growth and bulking and can be used instead of the traditionally practised removal of flowering shoots recommended by Heredia Zepeda (1977) and CCC applications can be used to increase the number of pods and the yield of seeds.

2.5 BIOCHEMICAL COMPOSITION

The edible parts of the tubers contain *ca.* 82–87% water, 0.7% fibre, 0.3–0.5% ash and 1.2–1.5% protein, 0.1–0.2% fat and 9.5–10.6% carbohydrate. They are a good source of vitamin C (0.2 mg/1000g) and also contain Fe 1.13, Ca 16.0, thiamin 0.5, riboflavin 0.02 and niacin 0.2 mg per 100 g respectively (Kay, 1973).

2.5.1 **Carbohydrates**

One of the complete determinations of the biochemical composition of yam bean (*P. erosus*) tubers was carried out by Tadera *et al.* (1984). Although the

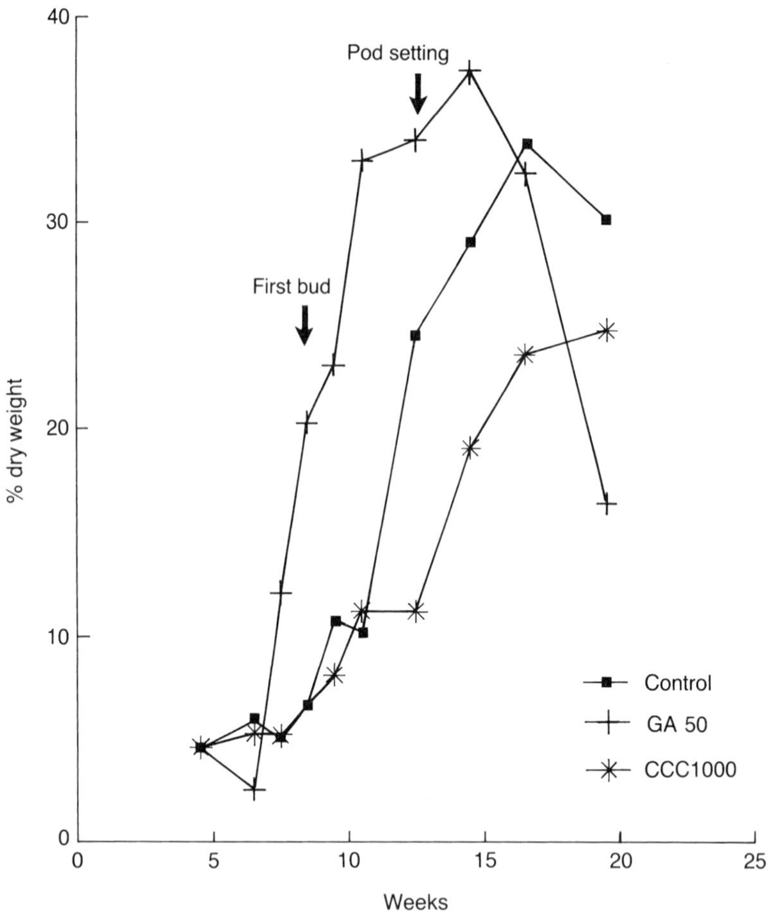

Figure 2.11 Development of total soluble sugars in the growing tubers of plants grown during short days and treated by chloride of chlormequat (CCC, 1000 p.p.m.) and gibberellic acid (GA, 50 p.p.m.).

tuber used was immature some of the results indicated the characteristics of the *P. erosus* tuber. However, results of biochemical analyses of other species within the genus *Pachyrhizus* are scarce.

The main reducing sugars are glucose and fructose and the main non-reducing one is sucrose. On the basis of fresh weight Paull *et al.* (1988) found little change in the composition during growth of the tuberous root. However, the study of sugar change in developing tubers showed that the soluble sugars in the control plants increased from the appearance of the first floral buds (R5 stage) up to the end of pod filling (beginning of R8 stage), then decreased. The starch increased in the tuber after the appearance of the first buds then the content decreased until the competition between pod filling and tuber bulking ceased (Figure 2.11).

Tubers of plants treated with GA had increased content of reducing sugars, whereas the content decreased when treated with CCC. Both growth factors favoured sucrose accumulation (Zinsou *et al.*, 1987b). GA induced an early accumulation of starch in the tuber with a peak accumulation at the appearance of the first floral buds (Figure 2.12).

The decrease in starch content observed in the control plants and in the CCC treated plants during the pod setting and filling disappeared, and confirmed the suppression of competition due to the abortion of the pods. The later increase in starch content coincides with the resuming of tuber growth during the R8 and R9 stages. Finally, Zinsou *et al.* (1987b and 1987c)

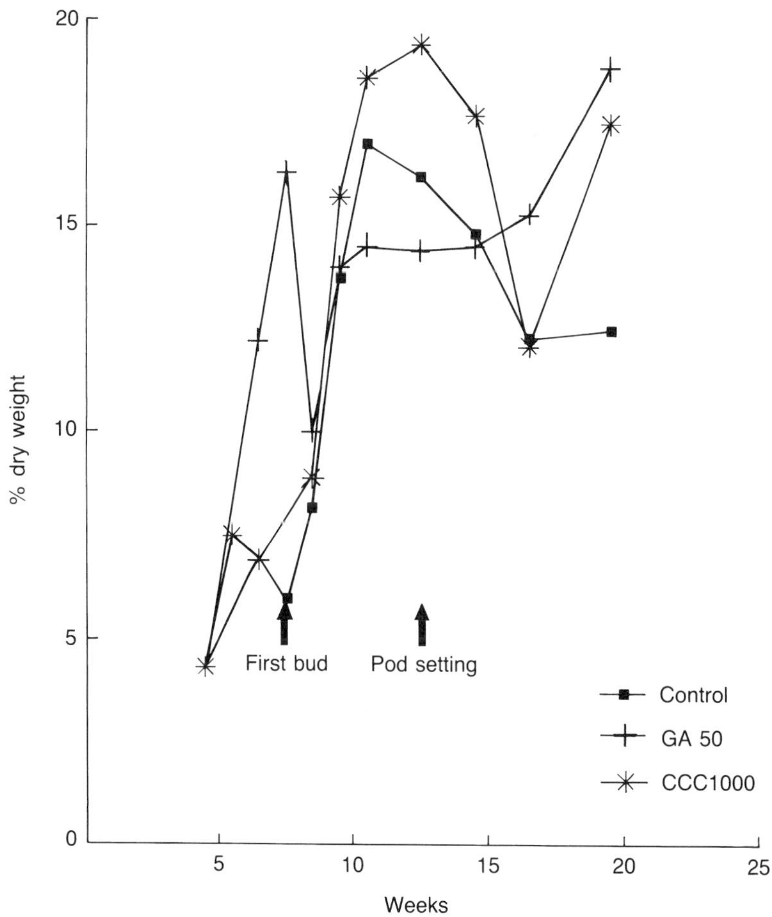

Figure 2.12 Development of starch content in the growing tubers of plants grown during short days and treated by chloride of chlormequat (CCC, 1000 p.p.m.) and gibberellic acid (GA, 50 p.p.m.)

demonstrated that treatment with both growth factors GA and/or CCC, will increase the starch content in the tubers at harvest.

Photomicrographs of yam bean (*P. erosus*) starch show that the starch is composed of spherical and compound granules with a mixture of small granules and comparatively larger ones (Tadera *et al.*, 1984). The average size is about 3.82 μm in yam bean compared to about 4.59 in winged bean. The X-ray patterns of *P. erosus* starch revealed a CA type different from the B type encountered in sweet potato (*Ipomea batata* (L.) Poir.) and Irish potato (*Solanum tuberosum* L.) starch. More than 80% of *P. erosus* starch was digested by glycoamylase within 24 hours. Tadera *et al.* (1984) explained the high digestibility of yam bean (*P. erosus*) starch by the small size of the granules and its low blue value scores in the starch - iodine test.

Nodules may reach a considerable size, up to 15 g fresh weight/nodule. This leads to the hypothesis that the host tissue conserves the ability to accumulate sugars as do the root tissues after tuber initiation.

Sugar composition of the nodule changes according to (a) the physiological stage of the plant, (b) the *Rhizobium* strain that nodulates the plant and (c) cultivation conditions/practices. Although glucose and fructose are present in the nodules until anthesis they disappear during pod setting and filling (Figure 2.13).

Starch is the major constituent through the vegetative stage of flowering. Generally the level is as high as 15–18% then it decreases and sucrose becomes the major constituent during pod filling. Figure 2.13 shows the change in the sugar composition in the nodules formed by two local (Guadeloupe) *Rhizobium* strains at four developmental stages of *P. erosus* (Chenevard, 1990). The presence of starch and sucrose in the nodules at pod filling indicates that the nodules are still functioning and supplied with assimilates.

2.5.2 **Amino acid and protein composition**

Srivastava *et al.* (1973) reported a protein content of 1.47 in fresh peeled tubers of *P. erosus*. Grum *et al.* (1990) analysed 8 cultivars of *P. erosus* and found a mean crude protein content (DM) of 8.3% (9.8–6.6%); four accessions of wild *P. erosus* material also yielded a mean of 8.3% (9.4–7.1%) and one accession of *P. ahipa* yielded 5.1% crude protein in the dry matter. Evans *et al.* (1977) analysed *P. erosus* tuber meal and found a total nitrogen content of 3.04% in the dry matter; 56% of the nitrogen encountered was in the amino-acids, peptides and proteins, 42.5% in non protein nitrogenous compounds. Tadera *et al.* (1984), using a different method of fractioning yam bean proteins, confirmed the results of Evans *et al.* (1977). The main proteins are albumins, globulins and glutelins.

The composition of amino-compounds in the nodules changes in relation to the developmental stages of the plants and the *Rhizobium* strains.

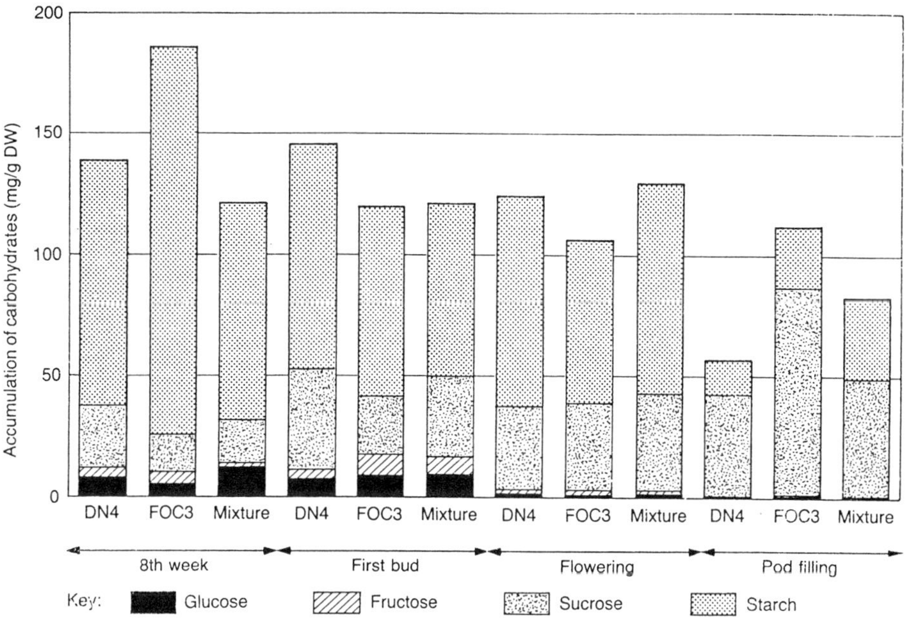

Figure 2.13 Accumulation of carbohydrates in the nodules formed from symbiosis between *Pachyrhizus erosus* (grown during long days) and two local strains of *Rhizobium*, DN4 and FOC3 and a mixture (both strains combined) at four key stages of plant growth: 8th week, first bud, flowering and pod filling.

Hence, the data presented here are only indicative. The soluble amino-compounds represent about 44% of the dry weight. The main proteins are globulins (6.2%). The other proteins are albumins (0.9%), prolamins (1.1%), glutelins (1.9%) and cell wall proteins (1.6%). Table 2.3 shows the evolution of the amino-compounds and the ureides in nodules formed from symbiosis of two local (Guadeloupe) strains of *Rhizobium* with *P. erosus* at four developmental stages. The amino-compounds other than the ureides are the major nitrogenous compounds. The ureides were found in low quantity at any stage of plant growth. This observation could be explained by the fact that after their formation in the nodules they are quickly transferred.

2.5.3 **Insecticide contents**

All the green parts of the plant possess insecticidal properties, but only the seeds have been used as an insecticide and a fish poison in tropical countries

Table 2.3 Composition in amino-compounds and in ureides of nodules formed from symbiosis between *Pachyrhizus erosus* and two indigenous Rhizobium strains and mixture at four different stages of growth during long days

	Amino-compounds (mg/g DW)			*Ureides (mg/g DW)*		
Stages	*DN4*	*FOC3*	*MIXTURE*	*DN4*	*FOC3*	*MIXTURE*
8th week	222	148	141	1.51	0.97	1.15
First bud	173 ± 38	301 ± 24	293 ± 62	1.39 ± 0.21	0.93 ± 0.05	0.87 ± 0.21
Flowering	64 ± 12	158 ± 60	117 ± 3	1.02 ± 0.26	1.64 ± 0.62	1.37 ± 0.27
Pod filling	64 ± 14	135 ± 37	69 ± 7	1.46 ± 0.6	2.63 ± 1.7	2.4 ± 1.7

DW, dry weight

(Liu and Hsu, 1941; Hansberry *et al.* 1947). Boorsma, (1910) discussed the toxicity of the seeds, leaves and roots of the plant to fish, dogs and other vertebrates. He called the toxic compound 'Pachyrhizid' by similarity of the toxic compound 'Derrid' extracted from *Derris* Lour. Pachyrhizid and pachyrhizone have formulae $C_{30}H_{18}O_8(OCH_3)$ and $C_{19}H_{12}O_6$, respectively (Shangraw and Lynn, 1955). The total extractive averaged 28% of the dry weight. Of this 90% was oil and the remainder was of a resinous nature.

Chemical analysis of the resinous fraction showed that the plant contained only a fraction of a percentage of rotenone. Therefore, the rotenone alone could not account fully for the toxicity of the plant (Norton, 1943; Hansberry and Lee, 1943). These authors suggested that the toxicity of *Pachyrhizus* was not primarily due to the rotenone content. Further investigations by Shangraw and Lynn (1955) demonstrated that the plant contained two pachysaponins A ($C_{30}H_{38}O_6$) and B ($C_{23}H_{18}O_7$) with respective melting points of 196–198°C and 176°C. The crude saponin mixture was found to be fatal to goldfish in a concentration of 1:10 000. From a complete review of the composition of the resinous fraction of the seeds according to the different authors it appears that ten closely related compounds exist. Krishnamurti *et al.*, (1970) suggested that the variation in the results observed in the different analyses published may be explained by seasonal and/or geographical effects. The different compounds are shown in Table 2.4. The role of the plant as a source of rotenone and other insecticidal compounds is disappointing (Norton and Hansberry, 1945).

2.5.4 **Others**

The seeds contain a fatty oil, somewhat similar to cotton seed oil. Fat content is > 25% and since protein content is of the same magnitude any future interest in extracting pesticidal compounds would lead to a useful residue for animal feed.

Table 2.4 The main insecticidal compounds isolated from the seeds of *Pachyrhizus*

Compound group	*Compound*
1 Isoflavone	Dehydroneotenone
1 Isoflavanone	Neotenone
6 Rotenoids	Rotenone
	Erosone
	Pachyrrhizone
	12α-OH pachyrrhizone
	Dolineone
	12α-OH dolineone
2 Furano-3-phenyl coumarins	Pachyrrhizin
	Erosnin

2.6 AGRONOMY, YIELD AND PRODUCTION

2.6.1 **Sowing and general agronomy (modified from Kay, 1973)**

Yam beans are mostly grown from seed but sprouted roots from a previous crop are also used and are favoured for clonal propagation of desirable genotypes.

Seeds, 2–3 per site, are placed in holes in well-prepared ridged beds and thinning is done as necessary. Spacing varies and can be 1–40 cm apart with 15–75 cm between rows but best results are obtained at *ca.* 15 × 15 cm spacing. 1000 seed weight is *ca.* 200 g.

Seed rates, if germination is high, varies from about 20 kg/ha to 45–50 kg/ha. The crop is kept weed free and usually mulched to keep weeds down. Yam beans, being climbing or trailing vines (except for *P. ahipa*) require trellises, and good support of the above-ground parts increases yields. Traditionally plants are stopped or pruned to encourage more vegetative growth and removal of flowers is often practised to increase yield and quality of tubers (see also Section 2.5).

Well-cultivated loams are optimal with proper drainage. The crop is not tolerant of water logging, but can be grown under irrigation. In traditional culture, compost is usually incorporated before planting. Yields are increased by fertilizers before planting (NPK 12:24:12 at 300–400 kg/ha) followed by ammonium sulphate at 200 kg/ha when plants begin to climb.

The crop usually matures in 5–8 months but depending on conditions, in hotter areas, maturation time may be as short as 3–4 months. For production of seed up to 10 months growth is required.

2.6.2 **Harvesting**

The fleshy roots may be solitary or several, simple or lobed and are turnip shaped. They are harvested when the crisp texture is optimal when they average 10–15 cm diameter and weigh over 2 kg. Larger tubers are tough and unappetizing.

Date of harvest is decided by the size of the tuber. Tuberous root development of *P. erosus* is sigmoidal (Paull *et al.*, 1988; Zinsou and Venthou-Dumaine, 1990). The dry matter content declines from 17.5% to 9% over 6 months (Paull *et al.*, 1988). There is little change in the quality characteristics of total sugars and acidity during growth. However, total phenols decline, though even the initial levels are very low and unlikely to impart astringency. Harvesting date based on tuber size would, therefore, be expected to have little effect on quality. The exception would be that younger tubers with higher dry matter percentages would be expected to have lower rates of water loss.

The tubers are either dug by hand or ploughed out. The same size tubers are braided or stacked together. After the tops are removed completely, the tubers are bagged or placed in field lugs or boxes. Tubers are then washed to remove dirt and sorted to remove diseased and unsaleable tubers before the tubers are taken to market in bags or boxes soon after harvest.

Grading and storage

The tubers should be free of dirt, mature, firm, and free from shrivel. In Hawaii, the preference is for a tuber that is flat on the top and bottom. Scalloping should be few in number and not too close together (Anon., 1974). Each tuber should not weigh less than 0.34 kg or more than 1.13 kg. The types of damage that are to be avoided are growth cracks, cuts, skin roughness, disease, insect damage and physiological breakdown, including chilling and freezing damage. Storage temperatures in the range of 0° to 5°C are reported to extend life (Bautista and Cadiz, 1967; Schroeder, 1967). However, yam bean tubers are chilling sensitive (Barile and Esguerra, 1984; Paull and Chen, 1988). Two weeks at 0°C leads to tubers showing pitting and softening. These tubers are regarded as inedible (Barile and Esguerra, 1984). The results of Barile and Esguerra (1984) suggest storage at 15°C based upon a storage study using 5°, 10°, 15° and 20°C. Weight loss and mould growth need to be controlled at 15°C. Other researchers have reported that 12.5°C can be used to extend postharvest life for at least a month (Paull and Chen, 1988). These and other unpublished results have led to the storage recommendation of 12.5°C to 17.5°C at 65 to 75% relative humidity (RH) for 1 to 2 months. The low RH recommendation, though it increases the rate of water loss, is used to reduce problems of superficial mould.

Postharvest technology

Postharvest problems include dehydration, fungal growth (see Section 2.6.4) and sprouting (Barile and Esguerra, 1984; Shroeder, 1967).

Tubers continually lose weight during storage (Paull and Chen, 1988). The rate of water loss is 0.12% of their initial weight per day when stored at 12.5°C and 0.21% per day at 22°C during the first month of storage (Figure 2.14a). Part of the difference is due to the different RH at different storage temperatures. After the first month of storage there is a reduction in the rate of water loss. This reduced rate may be due to the outer drier zone of the tuber acting as a barrier to further water loss. Yam bean water loss rates are considerably lower than the rates reported for other root and

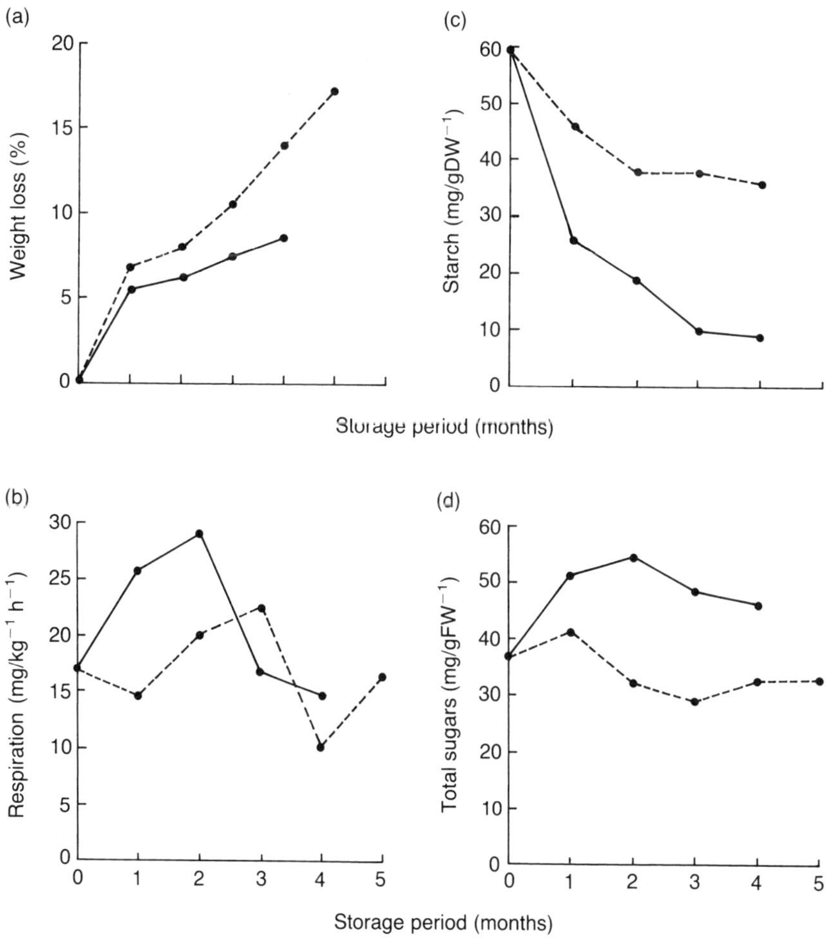

Figure 2.14 Changes over storage time, affecting (a) weight; (b) starch content; (c) respiration rate; (d) total sugar content.

tuber crops (Burton, 1973; Robinson *et al.*, 1975), except sweet potato and cassava (Singh *et al.*, 1952; Singh and Mathur, 1953).

The rate of respiration of yam bean tubers stored at 22°C is approximately 15 to 20 ml kg^{-1} h^{-1} for the first 2 months of storage, then declines to a level similar to tubers stored at 22°C (Figure 2.14 b). The increase in yam bean tuber respiration at 12.5°C was similar to the increase reported to occur in mature potato tubers stored at 2°C (Burton, 1973). This increase in respiration could be due to chilling injury in both tubers. Ethylene production has not been detected following sealing of the tuber in an air-tight jar for one hour, when the sensitivity of the assay is about 20 ml kg^{-1} h^{-1} (Paull and Chen, 1988).

The rise in respiration in tubers stored at 12.5% is related to the changes in starch (Figure 2.14 c), total soluble solids, total soluble sugars (Figure 2.14 d) and soluble sucrose, glucose and fructose (Paull and Chen, 1988). The level of starch declines to about two-thirds of the harvest value after 2 months' storage at 22°C (Figure 2.14 c). Tubers stored at 12.5°C have a considerably greater loss of starch, declining to one-third of the original value in 3 months. The initial peak of respiration for tubers stored at 12.5°C could be related to the increased starch breakdown shown by these tubers (Paull and Chen, 1988).

The breakdown of starch in tubers stored at 12.5°C led to total sugars increasing from 36 to 51 mg g^{-1} fresh weight (Figure 2.14 d), while sucrose increased from 950 to 2100 μg g^{-1} fresh weight. Sucrose levels in tubers stored at 22°C remain constant during the 4 months of storage. Glucose and fructose levels in the tubers continually decline under both storage conditions and did not correspond to the increase in sucrose (Paull and Chen, 1988). Similar changes in reserve carbohydrates and soluble sugars have been reported for potatoes stored at temperatures less than 5°C (Burton, 1966; Isherwood, 1976) and for sweet potatoes stored at 10° to 14°C (Scott and Matthews, 1957). Total soluble solids in yam bean tubers increase during the first month of storage at 22°C, then decline after 3 months (Figure 2.14 d). Total sugars show no significant changes during storage at 22°C.

During storage at 22° and at 12.5°C the amount of total phenols in the tuber gradually increased. The increase in total phenols in tubers stored at 22°C is more rapid and peaked after 2 months' storage (Paull and Chen, 1988). However, the overall level of total phenols is very low and therefore not likely to impart astringency to the tuber.

Titratable acidity (10 to 15 μeq g^{-1} fresh weight) remains constant during storage. Citric and malic acids are detected in small amounts (Paull and Chen, 1988).

Sprouting is observed only after extended storage of two to three months (Paull, unpublished). No sprouting is observed in tubers stored 4 weeks at 12.5°C (Paull, unpublished; Barile and Esguerra, 1984). There are no known reported studies on the use of sprout inhibitors.

Storage temperatures less than 15°C lead to loss of visual quality (Barile and Esguerra, 1984; Paull and Chen, 1988). These results suggest that yam bean tubers have a higher temperature injury threshold for chilling injury development than mature potatoes. Tubers can be stored for at least one to two months at 12.5°C (Paull and Chen, 1988), for 14 days at 10°C and for shorter periods at less than 10°C (Barile and Esguerra, 1984). These results suggest that the storage at temperatures less than 12.5°C does lead to chilling injury, an associated breakdown of starch and a loss of appearance. This loss of appearance can lead to downgrading of tubers in Hawaii (Anon., 1974).

Two types of Mexican *P. erosus* tubers have been described (Schroeder, 1967), the Guanajuato type with whitish peel and a translucent juice, and the Nayarit type with a milky or cloudy juice and brown peel. There was some indication of a change from the translucent to milky form upon storage at low temperatures (Schroeder, 1967). At 22° and 12.5°C, tuber flesh did develop an off-white colour following two months storage, however, no milky juice is observed (Paull and Chen, 1988).

2.6.3 **Yields**

Apart from qualitative results from plant introductions the crop has not been the subject of many yield trials. Nonetheless a study on genetic variation and correlations of different characters with yield is that by Sinha *et al.* (1977a and 1977b) involving 34 cultivars of *P. erosus* (Table 2.5).

2.6.4 **Pests, diseases, and spoiling**

In Central America the yam bean is attacked by larvae of *Theila jebus* and in Mexico the seeds are often attacked by weavils (Kay, 1973); on the whole, insect damage is slight.

Some foliar diseases occur: bacterial spot and two fungal diseases have caused damage in Mexico. In the Philippines a virus mosaic causing mottling or blistering of the leaves can affect tuber size negatively.

The only reported pathogenic fungus on tubers is *Rhizoctonia* root rot (Raabe *et al.*, 1981) in Hawaii. In the Philippines, severe *Penicillium* mould growth has been reported after storage at 0°, 5° and 15°C at greater than 90% RH (Barile and Esguerra, 1984). Considerable surface mould is observed on tubers stored at 12.5°C (Paull and Chen, 1988). Saprophytic surface mould (*Aspergillus* and *Fusarium*) formed cottony masses of mycelia that caused superficial damage. The recommendation of a 15°C storage minimum is probably necessary to avoid loss of tuber quality and to prevent mould development. Tuber rots became a problem after storage for one month at 22°C (Paull, unpublished.). A hypochlorite dip is sometimes used to reduce decay. The dip should maintain 55 to 70 p.p.m. available chlorine.

Table 2.5 Genetic properties of characters showing significant correlations with yield in *Pachyrhizus erosus*

Character		*Correlation*	*Mean*	*Heritability*
Days to first flowering	G	-0.5730**	90.65	95.00
	P	-0.7038		
Days to pod maturity	G	-0.5671**	199.04	83.33
	P	-0.7224		
Length of main stem (cm)	G	-0.6765**	106.09	85.42
	P	-0.7662		
Number of leaves/plant	G	-0.3468*	18.54	91.17
	P	-0.1893		
Number of pods/plant	G	-0.4121*	13.93	63.86
	P	-0.3981		
100-seed weight (g)	G	0.6414*	17.46	98.82
	P	0.7462		
100-seed weight (g)	G	0.6414**	17.46	98.82
	P	0.7462		
Root diameter (cm)	G	0.7858**	8.88	90.36
	P	0.7976		
Root length (cm)	G	-0.6577**	7.26	98.91
	P	-0.9185		
Yield/plant (g)			279.12	68.79

G, genotypic; P, phenotypic.
*significant at 5% level; **significant at 1% level. Only phenotypic correlations have been tested for significance. Source: Sinha *et al.*, 1977a and 1977b.

2.6.5 **Rhizobium**

There have been a number of studies on different aspects of nitrogen fixation in *Pachyrhizus erosus*. Woomer (1979) tested 23 strains of *Rhizobium* on *P. erosus* and found two that gave higher total nitrogen yields than a treatment with added nitrate. Vansuyt and Zinsou (1990) found in *P. erosus* that nitrogen fixation was poor on acid soils (pH 4.6). Lynd and Purcino (1987) found that P, Ca and K were all important for nitrogen fixation in *P. erosus*. Søren Kjaer (pers. comm., 1990) found that the efficiency of the nitrogen fixing symbiosis with *P. ahipa* was good in pot trials with vermiculite.

Before commencing new breeding programmes, it is important to decide whether to breed for symbiosis with the *Rhizobium* already existing in the soils of the region or whether to breed for symbiosis with a specific *Rhizobium* used for inoculation (Pulver *et al.*, 1985).

There is need for an international breeding programme aimed at low technology agriculture in particular localities, which ought to breed towards a 'promiscuous' host line that nodulates effectively and efficiently with a large range of strains. Such a line would be suitable for future breeding programmes to improve symbiosis of otherwise locally adapted varieties.

Sylvester-Bradley and Kipe-Nolt (1988) suggest a procedure for evaluating the legume-*Rhizobium* symbiosis as part of an overall legume selection programme. This procedure begins with the selection of parental material.

CIAT uses a recurrent selection programme and backcrossing for *Phaseolus vulgaris*, crossing the best nitrogen fixers with plants with a high yield potential (CIAT, 1982; Graham and Temple, 1984). The segregating generations are evaluated for dry matter accumulation under nitrogen-free conditions in the glasshouse. This method had a strong correlation ($r = 0.81$) with nodule dry weight in the field and a good correlation ($r = 0.45$) between glasshouse and field yields (CIAT, 1982). Different strains of *Rhizobium* are used in each cycle to limit specificity (Graham and Temple, 1984).

On the other hand, Mytton (1984) argues that it is impossible to eliminate all *Rhizobium* strain dependency in host genotypes and thus, the most rapid genetic advance would be obtained by coincidental selection of plant and *Rhizobium* genotypes.

2.7 GENETICS AND BREEDING

Until recently varieties of *Pachyrhizus* were largely primitive forms, although some selection in Mexico had led to named varieties 'Aguadulce' and 'Cristalina' released by Instituto Nacional de Investigaciones Agricolas, Celaya, Guanajuato. Also selection in *P. erosus* has led to the most common varieties being prostrate.

The self-fertilizing nature of the genus gives inbred homozygous lines as the input and output of breeding. This plays an important role in the statistics of inheritance and influences the methods and stages of selection.

The homozygosity means that no recessive alleles are hidden in the parental material, which makes for much easier evaluation of its value for breeding. The diploid ($2n = 22$) nature of the species (Section 2.2) makes it easier to evaluate recombining lines after a cross, as recessive alleles will show up in smaller groups of progeny than they would with higher levels of ploidy.

New cultivars of self-fertilized crops generally originate by introduction, selection, hybridization or a combination of these options. Sørensen (1988) gives several examples of introductions of *Pachyrhizus* into new areas. The introduction of *P. erosus* into Southeast Asia, probably by the Spanish, showed the most remarkable success. The lines of self-fertilizing crops grown without selection or collected from the wild are generally not pure homogeneous lines, but often heterogenous mixtures.

It is also clear that a limit to gain is reached when the best pure lines have been selected. In *P. erosus* pure line selection has been used for a number of years by plant breeder Ing. A. Heredia Zepeda in Mexico. When first used, this method approximately doubled yields and new, better yielding lines are still being selected in this way (Heredia Zepeda pers. comm., 1990). It is, however, expected that hybridization will now be necessary to maintain a high rate of gain.

It is possible, though, that superior parents for hybridization could be chosen amongst promising lines that are heterogeneous, particularly in the lesser researched species. Variation has been observed in many of the accessions screened in Tonga. The third possible origin of new varieties, the one on which most emphasis will be put here, is through those methods involving hybridization.

2.7.1 **Breeding options**

Since yam beans are seed propagated tuber crops the competition between seed and tuber production is significant. Hence, the economically best results of breeding would be obtained by the simultaneous selection for a number of viable seeds and tuber yield, weighing the two characters according to their relative economic values, thus obtaining a single value on which to base selection.

However, the long-term optimum ratio between seed and tuber could create difficulties during the breeding phase and early stages of seed multiplication, as each plant would produce a relatively small number of seeds. The problem would be aggravated if breeding was aimed at obtaining a smaller tuber through breeding for high density planting.

This makes it necessary to decide early in a breeding programme whether to select for a combination of the two characters or to select for a variety that gives a high tuber yield when flowers are picked.

Interspecific crosses have proven possible in *Pachyrhizus* spp. and appear to have great potential (see Section 2.2).

The chances are high that new combinations of genes from interspecific crosses will be inferior to the original ones and the hybrids will on average be less fit than either parental type (Allard, 1960). Large numbers are therefore needed to find transegregant plants. Nonetheless, an example of a promising interspecific cross in *Pachyrhizus* is a cross between a high yielding *P. erosus* and a *P. ahipa*. Yield in *P. erosus* has been found to be negatively correlated with the length of the main stem (Sinha *et al.*, 1977b). *P. ahipa* is generally very low yielding, but has a determinate growth habit, which may control the length of the main stem if the two are crossed and thus give rise to high yielding progeny. *P. ahipa* also appears to be insensitive to day length/temperature inhibition. Such crosses are in fact a major part of the ongoing breeding programme in Tonga.

2.7.2 **Selecting progeny**

Pedigree method

The classical selection procedure in populations after hybridization of self-fertilizing crops is the pedigree method. In this method selection begins in the F2 generation, the seeds of each F2 plant are grown separately and the best F3 plants chosen from the best families. This is continued in the following generations until the plants are assumed to be homozygous.

In this way inferior genotypes are discarded and desirable genotypes are able to recombine between each generation, possibly giving yet better combinations of genes.

In the generations following the F2 generation, the plants are selected on the basis of the performance of other progeny of the same F2 plant as well as the performance of the individual plant. This increases the reliability of selection.

Also, selection in each generation involves a different environment, which provides a good opportunity for genetic variability of important characters to be expressed and effective selection to be practised (Fehr, 1987).

A major disadvantage of the method is the amount of record keeping required (Fehr, 1987; Brim, 1966; Jensen, 1988). The method also requires more land and labour than other methods of inbreeding (Fehr, 1987). The effect of competition between plants is so great that evaluation in the early generations must be done in spaced plantings if individual plants are to be evaluated too. Particularly in interspecific crosses the differences in growth can be so large that some plants would totally cover other, possibly desirable, plants, if planted densely.

On the other hand, selection in spaced planting often does not reliably reflect the performance of the plant when grown in dense plantings. This becomes obvious when one considers that *Pachyrhizus* plants can vary from an erect upright type of less than 30 cm in height to a prostrate type of several meters length (Sørensen, 1988). A special problem arises when selecting for farming situations where inflorescences will be removed.

Bulk population method

In natural selection, which occurs in bulk breeding, less competitive plants that are maintained for long periods in the population are often found to be superior genotypes when grown in pure stands (e.g. barley, *Hordeum vulgare* L., Allard and Adams, 1969). This is explained by the greater reproductive capacity necessary for the less competitive plants to survive. High reproductive capacity is desirable in most self-fertilizing crops.

In *Pachyrhizus*, however, a high reproductive capacity would be expected to be at the expense of tuber growth. Population dynamics therefore seem to work against an increase in tuber yield and the bulk method, *per se*, would therefore appear unsuitable for *Pachyrhizus*.

Single seed descent

Single seed descent would allow for increased homozygosity, without favouring plants with a high reproductive capacity (Brim, 1966).

In this method each F2 plant is advanced by single seed descent until the desired level of homozygosity is reached. Selection for yield begins by bulking progeny that trace to a single F2 plant or by the pedigree method.

The main advantage of the method lies in the time that is saved in evaluation, harvesting and record keeping. This enables one to work with a larger number of crosses and more F2 plants, which is where most of the variation is present.

There is also the possibility of rapid generation advance, because methods that induce earlier flowering can be used.

Perhaps the biggest problem with the single seed descent method for *Pachyrhizus* will prove to be the continuous loss of plants in each generation, when seeds do not emerge, a plant is accidentally hoed or a plant is lost for other reasons. Martin *et al.* (1978) found that they lost 62%, 58% and 47% of the plants in three different populations grown at high densities for four generations. In interspecific crosses the problems with germination can be even greater.

As solutions to this problem, Fehr (1987) describes two methods in which a number of seeds are harvested from each plant; one in which the seeds from each plant are then grown in a hill planting and one in which the seeds are bulked and a sample used for generation advance.

Recurrent selection

Recurrent selection involves cycles of crossing and selection. The method has been most widely used in cross-fertilizing species, but has also found use in self-fertilizing species (Wricke and Weber, 1986). The main advantage of recurrent selection is that more recombination is permitted by repeated crosses, so undesirable genes are not fixed and can eventually be discarded. The major disadvantage is that the method is labour intensive, requiring large numbers of crosses.

The starting population may be the F2 generation of a single cross or the result of multiple crosses. The best lines from the population are then selected using any of the selection procedures discussed previously and then crossed again.

The low success rate achieved with cross-pollination in Tonga so far (10–15%) makes a recurrent selection programme difficult. However,

Sørensen (unpublished) has observed plants that do not open their pollen sacs and are therefore, functionally, male sterile. This would greatly facilitate the use of recurrent selection.

A special case of recurrent selection is backcrossing which may prove very useful in *Pachyrhizus* spp. for transferring characteristics from one species to another. Carpenter and Fehr (1986) used backcrossing to introgress genes from *Glycine soja* Sieb. et Zucc., a wild relative of soybean, into cultivated soybean, *G. max* (L.) Merr. They estimated that in the absence of selection at least three backcrosses to *G. max* were required to obtain an adequately high frequency of agronomically acceptable segregates for effective selection in a cultivar development programme. Some of the criteria they looked at, such as vining score, lodging and height, are relevant to *Pachyrhizus*, but these desirable characteristics do not today exist in one high yielding variety. The problem is therefore slightly different when carrying out interspecific crosses in *Pachyrhizus*. Selection between backcrosses could probably substitute for one or two backcrosses.

2.7.3 **Experimental techniques for *Pachyrhizus***

In relation to crossing, Sørensen (1989b) found that the pollen sacs would open significantly prior to the opening of the standard and wing petals and those of *P. tuberosus* would open as much as 36–48 hours before the petals.

No studies have been made on the proportion of possible natural cross pollination. However, field observations indicate that there may be some cross pollination in *P. ahipa* as the accessions of this species show more within-line variation than the other species do. *P. ahipa* flowers also have a more open structure.

In Tonga good results have been obtained by emasculating in the afternoon, when the stamens are long, and then pollinating the following morning. Emasculation must take place before the pollen sacs open and the flower must be covered to avoid cross pollination by insects and to prevent the stigma from drying out. The flowers may be covered with a bag and, after pollination, tape can be used around the flower.

The traditional way of coordinating flowering for crossing plants that flower at different times is to use sequential plantings during appropriate periods. Flowering can often be delayed or extended by pruning (Major, 1980) and removing the flowers that are not crossed.

Zinsou *et al.* (1987a) found that spraying with chloride and chlormequat (CCC) increased pod yield in *Pachyrhizus spp.*

In some cases cloning is useful. For instance, *P. ferrugineus* has been successfully cloned through cuttings of approximately three nodes each (Jorge Morera, pers. comm., 1990). The method has since been successfully used to multiply F1 seeds from a *P. ahipa* × *P. tuberosus* cross, during its vegetative phase, so that greater variation could be obtained in the F2.

Rapid generation advance

Søren Kjaer (pers. comm., 1990) found that *P. ahipa* inoculated with *Rhizobium* produced flowers two weeks before plants that had been supplied with nitrogen, when grown in a nitrogen-free medium. Phosphorus tends to hasten maturity (Major, 1980). In Tonga it has been found that *Pachyrhizus* seeds can be sown immediately upon completion of pod filling and require no period of dormancy.

If the day length/temperature sensitive plants are undesirable, then an F2 under long days may present an opportunity for selection by roguing the plants that grow long vines, before they totally cover the shorter plants.

Single-plant selection

At least some single-plant selection is necessary at some stage in all breeding programmes. This generally has to take place at a wider spacing to avoid the effects of intergenotypic competition and in the interest of getting good seed yields for the following generation. At this stage selection should be restricted to characters of high heritability (see Section 2.6.3). Day length/temperature sensitivity and growth habit would be relevant characteristics in *Pachyrhizus*. To improve evaluation of less heritable characteristics the field is often divided into blocks and the best plants chosen from each as suggested by Gardner (1961).

Heredia Zepeda (pers. comm., 1990) selects individual tubers in *P. erosus* plantings with commercial densities. The selected tubers are then planted aside in a spaced planting, where they can sprout and produce seeds. The advantage lies in the realistic conditions under which selection is made, but a year is lost for seed production. It must also be noted that the *P. erosus* material is relatively homogeneous, whereas interspecific crosses will have substantially more variation, making selection in dense plantings untenable.

Yield trials

England (1977) found that if replicates limited the number of families tested, then the advantage gained by the higher selection intensities obtained without replicates was greater than the loss of accuracy without replicates. Thus, what is a problem in *Pachyrhizus* can be turned to a virtue by increasing the number of crosses made and thereby the possible selection intensity.

When sufficient seeds become available to conduct replicated trials, information is maximized by placing the replicates in different environments (Sandison, 1959). This gives greater variation, but a more accurate prediction of future performance of the lines.

The extra work involved should be weighed against the long-term saving due to the increased selection intensity possible with a more accurate prediction of performance. Many programmes already conduct trials of later generations at several locations and can easily put in more plots. Different cultivation practices may be used as different environments (Jensen, 1988). For example, in Mexico *Pachyrhizus erosus* is both intercropped with maize and grown as a monocrop.

In the generations between single plant evaluation and yield trials, when there are still not enough seeds for a full size plot, it is advisable to plant the families in as big plots as possible rather than replicated hill plots, single row plots or other compromises. It is better to have one evaluation reflecting agricultural conditions than to replicate false information.

2.8 RESEARCH NEEDS AND FUTURE POTENTIAL

P. erosus has potential as a high yielding crop because of ease of production, high yield stability and adaptability. It has few pests and diseases. The variation in the genepool is also high with known potential for germplasm enhancement.

Wider use of the crop will relate heavily to consumer acceptability. Criteria have to be defined since the requirements of consumers are not clear. Schroeder (1967) states that the juice must be clear; Schneider (1986) finds the crisp, crunchy texture highly desirable in salads and cooked food. Paull and Chen (1988) mention that the phenol content could be undesirable if too high. In Tonga, tasters complained of an unpleasant smell in some cultivars. The high water content and easily digestible carbohydrates (Tadera *et al.*, 1984) make it a good snack (Schneider, 1986). However, when used in cooking, as the Chinese do, it may be altogether different criteria that are important.

In the USA, *P. erosus* is promoted as being a healthy food. Newsweek (July 30, 1990) cited Dr. Ornish as saying, 'The average American may find it hard to pass up a cheeseburger in favour of a jicama [*Pachyrhizus erosus*] - cucumber salad. But then the average American is in line for a heart attack'. Two separate goals for tuber quality are of special interest. One goal could be along the lines of the tuber type that is generally found in *P. erosus*, with a low dry matter content and suitable for eating raw. Another goal could be for a more starchy tuber suitable for cooking. This goal may prove to be more elusive. There is no certainty that the goal is attainable, but observations on the variation in dry matter content within the genus, 8–28% (Grum *et al.*, 1990) indicate that the possibility exists. The greatest attraction of this goal lies in the increase in demand that could result, if *Pachyrhizus* tubers could be used as a staple food which is both cheap and nutritious.

The characteristics that would make a *Pachyrhizus* tuber good for cooking have not been determined, but the most important trait is likely to be an increase in starch content. Dry matter content is generally a good initial estimate of starch content in tubers.

The best selection procedures for developing better cooking tubers in *Pachyrhizus* would probably be to test for high dry matter content in the early stages, either by cutting a wedge out of each tuber or using specific gravity. In later stages cooking tests (boiling, baking and frying) would probably be easier to use than a number of more complicated tests for measuring the parameters of cooking quality.

A number of general criteria for a 'good' tuber can probably also be found. A skin that is easy to pull off would be desirable to most consumers, as would an even, round shape. The flesh just under the skin is often a bit fibrous and is therefore frequently cut away (Schneider, 1986). Improvement of the layer just under the skin would therefore reduce waste.

There are already very high yielding varieties and tremendous variation in genetic properties. The genus also shows a wide range of adaptability and potential for further increasing adaptability through the incorporation of insensitivity to day length/temperature inhibition of tuber growth. Labour requirements for the crop could be decreased by producing plants with a higher tuber/seed ratio so that flowers do not need to be removed and with a more erect growth habit that would ease weeding.

As soon as possible, yield trials should be conducted using plot sizes large enough to reflect true agricultural practice. Thereafter, the trials should be replicated in as many localities as possible. There is a need for studies into the potential for improving nitrogen fixing capability under diverse soil and climatic conditions. An economic study of the effects of altering the ratio of seeds to tuber yield would greatly help define a direction for genetic improvement. Studies on the possibility of producing a variety with high starch content, suitable for cooking, deserve a special place in international breeding programmes.

In the end, it is the consumer that has the final say. Therefore, whether the consumer is the farmer himself or at a faraway market, a knowledge of his preferences is a prerequisite for a successful breeding programme and thus demands continued attention.

ACKNOWLEDGEMENTS

Thanks are due to the staff of the Botanical Section of the Department of Botany, Dendrology and Forest Genetics, Royal Veterinary and Agricultural University, Copenhagen; Dr Ib Friis, Botanical Museum, University of Copenhagen, Professor Ole Hamann, University of Copenhagen Botanical Garden and colleagues of the authors for helpful discussions.

REFERENCES

Allard, R.W. (1960) *Principles of Plant Breeding*, John Wiley and Sons, New York.

Allard, R.W. and Adams, J. (1969) Population studies in predominately self-pollinating species. *Amer. Naturalist*, **103**, 621–45.

Alvarenga, A.A. and Valio, I.F.M. (1989) Influence of temperature and photoperiod on flowering and tuberous root formation of *Pachyrhizus tuberosus*. *Ann. Bot.*, **64**, 411–14.

Anonymous (1974) Standards for Hawaii-grown yam bean roots. Div. of Marketing and Consumer Services, *Dept. of Agriculture State of Hawaii*, 3 pp.

Barile, T.V. and Esguerra, E.B. (1984) Low temperature storage of yam beans. Post Harvest Res. Notes, *Dept. of Horticulture, Univ. Philippines, Los Banos*, **1**, 23–5.

Bautista, O.D. and Cadiz, T.G. (1967) Yam bean, in *Vegetable Production in South East Asia*, (eds J.E. Knott and J.R. Deanon), University of the Philippines College of Agriculture Press, pp. 301–5.

Boorsma, W.G. (1910) Over de Vergiftigheid van de Bangkoewang-plant (*Pachyrrhizus angulatus* Rich.). *Teysmannia*, **21**, 624–37.

Brim, C.A. (1966) A modified pedigree method of selection in soybeans. *Crop Science*, **6**, 220.

Brücher, H. (1989) *Useful Plants of Neotropical Origin and Their Wild Relatives*, Springer-Verlag, Berlin.

Burton, W.G. (1966) *The Potato*, Veenman and Zonen, Wageningen.

Burton, W.G. (1973) Environmental requirements in store as determined by potential deterioration. *Proc. 7th Brit. Insect. Fungi Conf.*, pp. 1037–55.

Carpenter, J.A. and Fehr, W.R. (1986) Genetic variability for desirable agronomic traits in populations containing *Glycine soja* germplasm. *Crop Science*, **26**, 681–6.

Chenevard, D. (1990) Efficience et caractérisation biochimique de 2 souches de *Rhizobium* nodulant le *Pachyrhizus*. *Rapport de V.A.T. INRA Antilles-Guyane*, 13pp.

CIAT (1982) *Bean Program 1981 Report*, CIAT, Cali, Colombia.

Clausen, R.T. (1945) A botanical study of the Yam Beans (*Pachyrrhizus*). *Mem. Cornell Univ. Agric. Exp. Sta.*, **264**, 1–38.

England, F. (1977) Response to family selection based on replicated trials. *J. Agric. Sci.*, **88**, 127–34.

Evans, I.M., Boulter, D. Eaglesham, A.R.J. and Dart, P.J. (1977) Protein content and protein quality of tuberous roots of some legumes determined by chemical method. *Qual. Plant. Pl. Fds. Hum. Nutr.*, **27**, 275–85.

Fehr, W.R. (1987) *Principles of Cultivar Development. Vol. 1: Theory and Technique*, Macmillan, New York.

Fernandez, F., Gepts, P. and Lopez, M. (1986) Il programma de desarrollo de la planta de frijol (*Phaseolus vulgaris* L.). *CIAT*, Cali, Colombia.

Gardner, C.O. (1961) An evaluation of the effect of mass selection and seed irradiation with thermal neutrons on yield of corn. *Crop Science*, **1**, 241–5.

Graham, P.H. and Temple, S.R. (1984) Selection for improved nitrogen fixation in *Glycine max* (L.) Merr. and *Phaseolus vulgaris* L. *Plant and Soil*, **82**, 315–27.

Grum, M. (1990) *Breeding* Pachyrhizus *Rich. ex DC.: A review of goals and methods.* Ph.D. thesis. Dept. Crop Husbandry and Plant Breeding, Roy. Vet. Agric. Univ. Copenhagen, Denmark. 75 pp.

Grum, M., Stølen, O. and Sørensen, M. (1990) Yam bean (*Pachyrhizus* Rich. ex DC.) variety trials in Tonga, South Pacific: Fresh tuber yield, dry matter and nitrogen content. *Proc. 26th Ann. Meeting of the C.F.C.S., Puerto Rico; 29th July–4th August 1990*, 10 pp.

Hansberry, R. and Lee, C. (1943) The yam bean, *Pachyrrhizus erosus* Urban, as a possible insecticide. *J. Econ. Entomology*, **36**, 351–2.

Hansberry, R., Clausen, R.T. and Norton, L.B. (1947) Variations in chemical composition and insecticidal properties of the yam bean (*Pachyrhizus*). *J. Agric. Res.*, **74**, 55–64.
Heredia Zepeda, A. (1977) Effecto de desfloracion de la jicama (*Pachyrhizus erosus*) sobre el rendimiento. *Proc. Amer. Soc. Hort. Sci. (Tropical Regions)* **15**, 146–50.
Heredia Zepeda, A. (1985) Guia para cultivar Jícama en el Bajio. *SARH Folleto par productores* **15**, 1–11.
Hoof, W.C.H. van and Sørensen, M. (1989) *Pachyrhizus erosus* (L.) Urban, in *Plant Resources of South-East Asia* (eds E. Westphal and P.C.M. Jansen), *Pudoc*, Wageningen, pp. 213–15.
Ingham, J.L. (1979) Isoflavonoid phytoalexins of yam bean (*Pachyrrhizus erosus*). *Z. Naturforsch.*, **34c**, 683–8.
Isherwood, F.A. (1976) Starch-sugar interconversions in *Solanum tuberosum*. *Phytochem.*, **12**, 2579–91.
Jensen, N.F. (1988) *Plant Breeding Methodology*, John Wiley and Sons, New York.
Kay, D.E. (1973) Yam bean. Potato bean, in *Root Crops* (ed. D.E. Kay), Tropical Products Institute, London, pp. 240–5.
Krishnamurti, M., Sambihy, Y.R. and Sheshadri, J.R. (1970) Chemical study of Indian yam beans (*Pachyrhizus erosus*), isolation of two new rotenoids: 12a-hyroxydolineone and 12a-hydroxypachyrhizone. *Tetrahedron*, **26**, 3023–7.
Lackey, J.A. (1977) A revised classification of the tribe Phaseoleae (Leguminosae: Papilionoideae), and its relation to canavanine distribution. *Bot. J. Linn. Soc.*, **74**, 163–78.
Lackey, J.A. (1981) Phaseoleae in *Advances in Legume Systematics, Part 1*, (eds R.M. Polhill and P.H. Raven), H.M.S.O., Kew, U.K., pp. 301–28.
Lamaze, T., Vansuyt, G. and Zinsou, C. (1985) Répartition des formes d'azote et catabolisme des purines chez le dolique tubéreux, *Pachyrhizus erosus* Urban. *Agronomie*, **5**, 933–8.
Liu, C.Y. and Hsu, Y.F. (1941) Studies on a new vegetable insecticide, the yam bean (*Pachyrhizus erosus* Urban), *Kwangsi Agriculture*, **2**, 28–48 (In Chinese).
Lynd, J.Q. and Purcino, A.A.C. (1987) Effects of soil fertility on growth, tuber yield, nodulation and nitrogen fixation of yam bean (*Pachyrhizus erosus* (L.) Urban) grown on a typic eutrutox. *J. Plant Nutr.*, **10**, 485–500.
Major, D.J. (1980) Environmental effects on flowering, in *Hybridization of Crop plants*, (eds W.R. Fehr and H.H. Hadley), American Society of Agronomy, Madison, pp. 1–15.
Martin, R.J., Wilcox, J.R. and Laviolette, F.A. (1978) Variability in soybean progenies developed by single seed descent at two plant populations. *Crop Science*, **18**, 359–63.
Martínez, M. (1979) *Catálogo de Nombres Vulgares y Científicos de Plantas Mexicanas*. Fondo de Cultura Económica, Mexico.
Masefield, G.B. (1957) The nodulation of annual leguminous crops in Malaya. *Emp. J. Exper. Agric.*, **25**, 139–50.
Mytton, L.R. (1984) Developing a breeding strategy to exploit quantitative variation in symbiotic nitrogen fixation. *Plant and Soil*, **82**, 329–35.
National Research Council (1979) *Tropical Legumes: Resources for the Future*, National Academy of Sciences, Washington, D.C.
Norton, L.B. (1943) Rotenone in the yam bean (*Pachyrrhizus erosus*). *J. Amer. Chem. Soc.*, **65**, 2259–60.
Norton, L.B. and Hansberry, K. (1945) Constituents of the insecticidal resin of the yam bean, *Pachyrrhizus erosus*, *J. Amer. Chem. Soc.*, **67**, 1609–14.
Pate, J.S., Atkins, C.A. White, R.J. *et al.* (1980) Nitrogen nutrition and xylem transport of nitrogen in ureide producing grain legumes. *Plant Physiol.*, **65**, 961–5.

Paull, R.E. and Chen, N.J. (1988) Compositional changes in yam bean during storage. *HortSci.*, **23**, 194–6.
Paull, R.E., Chen, N.J. and Fukuda, S.K. (1988) Planting dates related to tuberous root yield, vine length and quality attributes of yam bean. *HortSci.*, **23**, (2), 326–9.
Porterfield, W.M. (1939) The yam bean as a source of food in China. *N.Y.J. Bot. Gard.*, **40**, 107–8.
Pulver, E.L., Kueneman, E.A. and Ranga Rao, V. (1985) Promiscuous nodulating soybean efficient in N_2 fixation. *Crop Science*, **25**, 660–3.
Raabe, R.D., Conners, I.L. and Martinez, A.P. (1981) Checklist of plant diseases in Hawaii. Hawaii Institute of Tropical Agriculture and Human Resources, University of Hawaii. *Information Text Series*, **22**, 1-313.
Robin, C., Vaillant, V., Vansuyt, F. and Zinsou, C. (1990) Assimilate partitioning in *Pachyrhizus erosus* during long-day vegetative development. *Plant Physiol. Biochem.*, **28**, 343–9.
Robinson, J.E., Browne, K.M. and Burton, W.G. (1975) Storage characteristic of some vegetable and soft fruits. *Ann. Appl. Biol.*, **81**, 399–408.
Sandison, A. (1959) Influence of site and season on agricultural variety trials. *Nature*, **184**, 834.
Schneider, E. (1986) Jícama, in *Uncommon Fruits and Vegetables: a Commonsense Guide*, Harper and Row, New York, pp. 240–4.
Schroeder, C.A. (1967) The jícama, a rootcrop from Mexico. *Proc. Amer. Soc. Hort. Sci., (Tropical Regions)*, **11**, 65–71.
Scott, L.E. and Matthews, W.A., (1957) Carbohydrate changes in sweet potatoes during curing and storage. *Proc. Amer. Soc. Hort. Sci.*, **70**, 407–18.
Shangraw, R.F. and Lynn, E.V. (1955) The saponins of *Pachyrrhizus erosus* (L.) Urban. *J. Amer. Pharm. Assoc.*, **44**, 38–9.
Singh, K.K., Kapur, N.S. and Mathur, P.B. (1952) Studies in cold storage of beet roots, cabbages, cauliflowers and sweet potatoes. *Ind. J. Hort.*, **1**, 1–12.
Singh, K.K. and Mathur, P.B. (1953) Cold storage of tapioca roots. *Bull. Cent. Fd. Technol. Res. Inst., Mysore*, **2**, 181–2.
Sinha, R.P., Prakash, R. and Haque, P. (1977a) Genetic variability in yam bean (*Pachyrhizus erosus* Urban). *Tropical Grain Legume Bulletin*, **7**, 21–3.
Sinha, R.P., Prakash, R. and Haque, P. (1977b) Genotypic and phenotypic correlation in yam bean (*Pachyrrhizus erosus*). *Tropical Grain Legume Bulletin*, **7**, 24–5.
Sørensen, M. (1988) A taxonomic revision of the genus *Pachyrhizus* Rich. ex DC. *nom. cons. Nord. J. Bot.*, **8**, 167–92.
Sørensen, M. (1989a) Pollen morphology of species and interspecific hybrids in *Pachyrhizus* Rich. ex DC. (Fabaceae: Phaseoleae). *Review of Palaeobotany and Palynology*, **61**, 319–39.
Sørensen, M. (1989b) Cross breeding experiments: compatibility, pollen fertility, and germination percentages of interspecific hybrids, F1, F2 & F3 in the genus *Pachyrhizus* Rich. ex DC. (Fabaceae: Phaseoleae). *Proc. 25th Annual Meeting of the C.F.C.S., Guadeloupe, 2nd–8th July, 1989*, 39 pp.
Sørensen, M. (1990) Observations on distribution, ecology and cultivation of the tuber-bearing legume genus *Pachyrhizus* Rich. ex DC. (Fabaceae: Phaseoleae). *Wageningen Agricultural University Papers*, **90**(3), 1–38.
Srivastava, G.S., Shukla, G.S. and Awaasthi, D.N. (1973) We can grow sankalu in the plains of Uttar Pradesh. *Indian Farming*, **23**, 2.
Standley, P.C. (1930) Flora of Yucatan. *Publ. Field Mus. Nat. Hist., Bot. Ser. 279*, **3**, 157–492.
Sylvester-Bradley, R. and Kipe-Nolt, J. (1988) *The Legume-Rhizobium Symbiosis: Evaluation, Selection and Agronomic Management*, CIAT. Cali, Colombia.

Tadera, K., Tanguchi, T., Teremoto, M. *et al.* (1984) Protein and starch in tubers of winged bean, *Psophocarpus tetragonolobus* (L.) DC. and yam bean *Pachyrhizus erosus* (L.) Urban. *Mem. Fac. Agric. Kagoshima Univ.*, **20**, 73–81.

Urbina, M. (1906) Raices comestibles entre los antiguos Mexicanos. *Anales Mus. Nac. Mexico*, **3**, (ser. 2), 117–90.

Vaillant, V., Robin, C. and Zinsou, C. (1990) Effects of inhibition of nodule metabolism by nitrate on the regulation of assimilate partitioning in *Pachyrhizus. Plant Physiol. Biochem.*, **28**, 131–6.

Vaillant, V., Robin, C. and Zinsou, C. (1991) Comparison of the effects of allopurinol and nitrate on photosynthate partitioning in yam bean, in *Proc. 25th Ann. Meeting of the Caribbean Food Crops Society*, Guadeloupe, 1–6 July 1989, (ed. L. Degras), Publication INRA Antilles-Guyane, Guadeloupe, pp. 458–65.

Vansuyt, G. and Zinsou, C. (1991) Preliminary investigations on indigenous strains of Rhizobium nodulating yam bean (*Pachyrhizus erosus* (L.) Urban) in Guadeloupe, in *Proc. 25th Ann. Meeting of the C.F.C.S., Guadeloupe, 1–6 July 1989*, (ed. L. Degras), Publication INRA Antilles-Guyane, Guadeloupe, pp. 497–507.

Woomer, P.L. (1979) *Root tuberization and nitrogen fixation by* Pachyrhizus erosus *(L.) Urban*, M.Sc. thesis, University of Hawaii.

Wricke, G. and Weber, W.E. (1986) *Quantitative Genetics and Selection in Plant Breeding*, Walter de Gruyter, Berlin.

Zengbé, M. and Salsac, L. (1983) Variation des teneurs en uréides et en enzymes du catabolisme purique chez *Glycine max. Physiol. Vég.* **21**, 67–76.

Zinsou, C. (1988) Le dolique tubéreux (*Pachyrhizus erosus* Urban): Un nouveau légume pour les Antilles. *Bull. Agron. Antilles-Guyane*, **6**, 42–8.

Zinsou, C. and Vansuyt, G. (1991) Consequences of purine degradation inhibition induced by allopurinol on growth, nitrogenous compounds and carbohydrates in yam bean (*Pachyrhizus erosus* Urban), in *Proc. 25th Ann. Meeting of the C.F.C.S., Guadeloupe, 1–6 July 1989*, Publication INRA Antilles-Guyane, Guadeloupe, pp. 448–57.

Zinsou, C. and Venthou-Dumaine, A. G. (1990) Effects of different sowing dates on the growth and development of yam bean (*Pachyrhizus erosus* (L.) Urban). *Proc. 24th Ann. Meeting of the C.F.C.S., Jamaica, 15th–20th August 1988*, pp. 156–62.

Zinsou, C., Venthou-Dumaine, A. and Vansuyt, G. (1987a) Croissance et développement du *Pachyrhizus erosus* Urban. I Effets de l'acide gibbérellique et du chlorure de chlorocholine en jours cours. *Agronomie*, **7**, 677–83.

Zinsou, C., Vansuyt, G. and Venthou-Domaine, A. (1987b) Croissance et développement du *Pachyrhizus erosus* Urban. II. Effects de l'acide gibbérellique et du chlorure de chlorocholine sur la composition glucidique du tubercule. *Agronomie*, **7**, 821–5.

Zinsou, C., Vansuyt, G. and Venthou-Dumaine, A. (1987c) Gibberellic acid and CCC changed sugar composition and tuber yield in yam bean, *Pachyrhizus erosus* Urban. *Proc. 23rd Ann. Meeting of the C.F.C.S./C.A.E.S., Antigua, 24th–28th August 1987.*

3

Lupins (*Lupinus* species)

N. Haq

3.1 HISTORY AND USES

Lupins play an important role not only for agricultural development in diverse cropping systems but also they have a long history of use in human and animal nutrition. As with many leguminous species, traditionally related to smallholder farming, they can be multi-purpose plants. Most importantly they can play an important part in maintaining soil fertility or as an aid in restoring degraded soils, in particular poor soils and somewhat dry climates.

The seeds, as grain legumes, have been considered of greater potential use, because their protein content can be among the highest of any legume seed.

There is a wide range of species which appear to have potential for cultivation. They are distributed both in the Old World, largely the Mediterranean and African regions and in the Americas. There are four major domesticates: *Lupinus albus* (white lupin, field lupin, tremoco or wolf bean), *L. luteus* (yellow lupin) and *L. angustifolius* (European blue lupin) from the Old World and *L. mutabilis* (tarwi, pearl lupin, chocho, tarhui or taura) from the New.

The Egyptian lupin, which became widely cultivated in Egypt and Italy, *L. termis*, is now regarded as a type of *L. albus*, the species probably most cultivated in the Mediterranean in the past.

Whereas *L. albus* diversified into many local cultivars and its wild progenitor is well-known (*L. albus* subsp. *graecus*) native in the Aegean, *L. luteus* and *L. angustifolius* represent a continuum of forms from wild to cultivated. Cultivated *L. albus* has been recorded in archaeological sites in Bronze Age Cyprus and Roman Egypt and wild forms in the Natufian beds in Israel from the pre-farming Mesolithic period (Zohary and Hopf, 1988).

L. mutabilis was domesticated by pre-Inca people at least 1500 years ago

Pulses and Vegetables. Edited by J. T. Williams. Published by Chapman & Hall,
2–6 Boundary Row, London SE1 8HN. ISBN 0 412 46610 4.

and it became a significant contributor to the Andean region of Peru, Bolivia, Ecuador and Chile (NAS, 1979). This species has remained localized, but sweet lupins from the Mediterranean have received most attention and have been cultivated in Ethiopia, USA, South Africa, Australia, Eastern Europe, Chile and New Zealand.

Seeds of lupins are bitter-tasting, due to the presence of toxic alkaloids. R. von Stengbusch, in the 1920s and 1930s in Germany, isolated low-alkaloid strains. These became the 'sweet' forms. In the 1970s attention was paid to alkaloids in *L. mutabilis* and the selection of low-alkaloid forms (von Bauer and Gross, 1977, 1978).

A number of other species have been variously gathered: seeds of *L. perennis* in N. America were widely eaten last century but no longer. However, throughout history, the Old World lupins were associated with poor people. Additionally there was the belief that plants can 'take over' the land hence the name from Latin *lupus*, a wolf. Lupins are widely discussed in early literature of Egypt, Greece and Rome. Whereas the seeds were used to make a flour for bread, seeds were also, because of their bitterness, used for medicinal purposes. Lupin meal use for human food was widespread in large areas of Spain, Italy and Corsica until this century. Meal and flour were prepared from dried, ground and cooked seeds and the bitter taste removed by washing. A limited amount of washed, boiled whole seeds are still eaten as a snack food in the Mediterranean. Tarwi seeds are also washed for several days to leach out toxins. Not all species produce seeds with toxins; many of the species can be crossed as there are great opportunities for improvement through hybridization.

Scientific interest has focused on the two regions of origin of domesticates and a series of International Lupin Conferences have met over the past decade or so in parallel with a series of International Congresses devoted to Andean cultivated plants. Tarwi was singled out as a resource for the future in a publication on tropical legumes (NAS, 1979) and research strengthened in Peru. Also research interest in Europe has accelerated on the possibility of lupins as alternative crops and substantial projects have looked at this in the UK and France. Other long-established research in Eastern Europe has continued, especially in Poland and work in Germany and Western Australia continued. There is, as a result, a substantial research base, much of which is summarized in this account. Nevertheless, the account will show important gaps in our current knowledge.

Germplasm has also received heightened attention and a degree of gap-filling for collections was sponsored by the International Board for Plant Genetic Resources (IBPGR) and the EEC in southern Europe in the 1980s and funding mobilized for collecting the lupin genepool in the Andes.

The interest, earlier this century, in lupins related to their use as cover crops especially for soils of low fertility following widespread use for this purpose in the seventeenth and eighteenth centuries. Recently a wild lupin species in the Pacific Northwest of the USA, *L. albicaulis*, has been domesticated for this purpose and another, *L. nootkatensis* is being examined for this purpose in Alaska (Rumbaugh, 1990). Green industry and land reclamation were major considerations. They became of interest for animal feed, once the sweet forms had been developed. All aerial parts of the plant are toxic although the sweet forms of *L. albus* can be used for forage and lupins can produce lupinosis in animals. Hence interest focused on use of seeds in poultry, pig, sheep and cattle feed in a similar way to the use of soyabean meal. Less attention has been paid to expanded use as a human food, in part because the primitive cultivars have strong ecological requirements in relation to soil type and temperature regimes. Wider adaptability is possible through better screening of germplasm and crossing – even between distantly-related species. Current research is, in effect, providing a second wave of domestication. For a long time, seeds have been eaten in Ethiopia and Sudan and other poor countries in times of food scarcity. There are, in many countries, problem agro-ecologies where lupins find a niche, e.g. in the Sudan they grow on flooded lands that are too heavy or saline for other crops (Whyte *et al.*, 1953) and major international emphasis on sustainability and careful husbanding of more marginal areas can provide opportunities for lupins to be used much more widely as a supplemental human food. There are also much wider opportunities to process meal since most current soya products can be produced from lupin seeds.

3.2 TAXONOMY AND ECOLOGY

The genus *Lupinus* belongs to the tribe Genisteae of the Leguminosae. Parts of the genus have been the subject of revision, notably for the Mediterranean species by Gladstones (1974), for American species by Dunn (1984) and Planchaelo (1985), and the genus was revised by Plitman and Hayn (1984) on the basis of biological species isolated from one another.

Species may be herbs, rarely shrubs and annual or perennial, with well developed taproot and root systems. Stems are coarse and thick and usually do not branch until the commencement of flowering. Leaves are divided palmately with 5–15 leaflets (rarely 1–3). Leaflets are entire and linear to oblanceolate. Leaves are stipulate with stipules adnate to the base of the petiole. Petioles and stems are usually hairy. Flowers are large and vary in colour from white to yellow, blue, violet or variegated. Flowers may be scattered, or in cohorts in erect spikes. Flowers are typically papilionate with a deeply cleft calyx with persistent bracteoles adnate to the calyx base. The calyx is deeply cleft. The pod is usually compressed, hairy, and septate between the seeds and the valves are tough. Most are dehiscent except in many cultivars, although 'shattering' remains in many primitive cultivars of Old World cultigens. Lupins are usually self-fertilized, but cross-pollination can occur.

Table 3.1 Distribution and chromosome numbers of the cultivated species of lupins and relatives

Species	*Section*	*Distribution*	*Chromosome number (22)*	*Status*
I Old World				
Lupinus albus L. Syn. L. *graecus* Boiss. Sprun.	Albus	Greece, Yugoslavia, W. Turkey	30,40,48,50	C/W
L. angustifolius L.	Angustifolius	Pan-Mediterranean	40,48	C/W
L. luteus L.	Luteus	W. Mediterranean E. Mediterranean	46,48,50,52,104	C/W
L. hispanicus Boiss. Reut.		Iberia	52	W
L. micranthus Guss.	Micranthus	Pan-Mediterranean	48,52	W
L. cosentinii Guss.		W. Mediterranean	32	W
L. pilosus Murray		E. Mediterranean	42, 50	W
L. palestinus Boiss.		S.E. Mediterranean	42	W
L. digitatus Forsskal		Nile Valley and Sahara	36,40,42,52	W
L. princei Harms		E. Africa	?	W
L. atlanticus Gladst.		Morocco	38	W
L. somaliensis Baker f.		N. Somalia	?	W
II New World				
L. mutabilis Sweet		S. American Highlands (Andes)	42,48	C

From Lopez-Bellido and Fuentes, 1986; Smartt, 1990
C, cultivated; W, wild.

Early classifications emphasized the presence or absence of bracteoles attached to the calyx for distinguishing species especially related to *L. albus*, but is no longer a major diagnostic character. Domesticated species and relatives are easily distinguished (Tables 3.1 and 3.2); illustrations are clear in Maire, (1987) and Thulin (1983).

Chromosome numbers vary between and within species, an unusual situation in legumes (Table 3.1). Darlington and Wylie (1955) suggested a basic chromosome number $x = 12$ which suggests that the evolution of the species has entailed polyploidy, structural changes and perhaps aneuploidy. More detailed cytological studies on the species would eliminate some of the current taxonomic uncertainties within the genus. Gladstones (1984) placed the Old World species into five sections of the genus (Table 3.3). Section Albus consists of one species with cultivated and wild varieties, sometimes regarded as subspecies. Section Angustifolius consists of one species although a number of synonyms reported in the literature has caused some confusion and there may be a distinct subspecies *reticulatus*

Table 3.2 Some morphological characters of cultivated species of lupins

Species	Habit	Plant height (cm)	Seed					Flower		Fruit
			Surface	Shape	Size (mm)	Colour	weight (mg)	Colour	arrangement on inflorescence	length × width (mm)
L. albus	Short, hairy annual	30–120	Smooth	Square compressed	8–14	Dull, light yellow	360	White, tinged with violet or blue	Spiral	70–150 × 12–20
L. luteus	Short, hairy annual	25–80	Smooth	Orbicular quadrangular compressed	6–8	Tall, black marbled white	130	Yellow	Verticillate	40–60 × 10–14
L. angustfolius	Short, hairy annual	20–150	Smooth	Globular	6–8	Yellow brown, dark brown, or grey with yellow spots	170	Normally blue but white in cultivars	Spiral	35–50 × 7–10
L. mutabilis	Annual	Normally 50–180 but ranging up to 250	Smooth	Ovoid, compressed	6–10	Black, brownish black, white or white with black or grey around hilum	200	Purple, blue, Pink, white	Partially verticillate	70–100 × 18–24

Adapted from Duke (1981), Pate, Williams and Farrington (1985)

Table 3.3 Groupings of cultivated and related Old World lupins

Section	*Taxa*	*Natural distribution*
Albus	*L. albus*	
	var. *albus*	cultivated
	var. *graecus* Boiss. and Spruner	S. Balkans, Aegean, W. Turkey
Angustifolius	*L. angustifolius*	cultivated Pan-Mediterranean
	subsp. *reticulatus* (Desv.)	Arcangeli
Luteus	*L. luteus*	
	subsp. *luteus*	W. Mediterranean
	subsp. *orientalis*	E. Mediterranean
	L. hispanicus	
	subsp. *hispanicus*	C. and W. Spain
	subsp. *bicolor* Merino	Spain, Portugal
Micranthus	*L. micranthus*	Pan-Mediterranean
Rough-seeded	*L. pilosus*	E. Mediterranean
	L. palestinus	S.E. Mediterranean
	L. cosentinii	W. Mediterranean
	L. atlanticus	S. Morocco
	L. princei	E. Africa
	L. somaliensis	N. Somalia

which is wild. Section Luteus consists of two species, cultivated *L. luteus* and wild *L. hispanicus*. Section Micranthus is little known and consists of only one wild species *L. micranthus*. The 'Rough Seeded' Section consists of some related wild species.

The four cultigens are associated with agroecosystems, mostly on poor soils. *L. albus* and *L. angustifolius* are also associated with anthropic landscapes but wild forms of the latter can be found in grasslands, scrubland and thickets and woodlands. *L. luteus* appears to be associated with grasslands.

In more northern latitudes lupins are grown in the warmest season of the year but in the subtropics in the coldest season. All Old World cultigens are long-day plants, *L. mutabilis* is a short-day plant in the Andes but flowers in long days in temperate zones.

L. albus is widely cultivated as a secondary legume throughout the Mediterranean region, the Upper Nile, the Canary Islands, Portugal, Madeira, Central and Southeastern Europe, the former USSR, Ethiopia, South Africa, Australia, the Southeastern USA and South America (Table 3.3). It is well adapted on well drained alluvial, acid and slightly calcareous sandy loams. *L. albus* responds to vernalization and requires at least a five-month period with monthly temperatures 15–15°C and rainfall distribution between 400–1000 mm per annum. *L. albus* is fairly tolerant to frost. The pH of soils recorded are in the range 4.8–8.2 (average 6.4).

Germplasm accessions have been examined and their patterns of geographical variation analysed (Simpson, 1986a, 1986b). *L. albus* was introduced to new areas as Southern European migrants moved in the nineteenth century. It was introduced to Portugal and the former USSR at the beginning of this century.

L. angustifolius is cultivated throughout the Mediterranean and in Southwest France. It is now grown in Australia, New Zealand, South Africa, Northern Europe and Southeast USA. *L. angustifolius* grows on well-drained soils, neutral or moderately acid (pH 4.9–8.2 but average 6.6) loams or sandy loams and requires strong vernalization. It is better adapted to relatively cool growing conditions and has a very strong tendency to flower abscission at high temperatures. The species is also fairly resistant to frost. The annual temperature for growth is 5.6–26.2°C and the rainfall is 450–1000 mm per annum.

L. luteus is cultivated in West Spain and Portugal, coastal areas around the Mediterranean basin, and also in Northern Europe, South Africa and Australia. Flowering is influenced both by photoperiod and vernalization. It tolerates only light frost and requires mild growing season temperatures. It grows in acid and sandy loam soils and requires 450–1000 mm rainfall per annum, temperatures 6–26°C (mean 13°C) and soil pH 4.5–8.2 (mean 6.5). It can be grown at high altitudes (2000 m).

L. mutabilis is mainly cultivated in the Andes at high altitudes (1800–4000 m) and at low latitudes (0–22°S) but it is also being grown on a trial basis in Europe. It grows on acid sandy soils and is tolerant to drought and frost. The crop has been experimented on in Essex and in Hampshire in the UK under rainfed conditions and has been found to be suitable for this region (Chaudhary and Haq, unpublished.). The ecological requirements of the four cultigens are summarized in Table 3.4.

3.3 BIOCHEMICAL COMPOSITION AND PLANT ANALYSIS

The primary product of all four species is seeds. These contain 28.0–45.9% crude protein and two species, *L. albus* and *L. mutabilis*, contain 11% and 13–23% oil respectively (Table 3.5). The high protein content in the seeds of these species makes them important in protein deficiency situations for human and also animal nutrition. The Andean species shows the highest potential for protein and oil sources, with a similar range of protein and oil content to soya and the crop can be grown in areas where soyabean cannot be grown (Williams 1979). Apart from a ready source of high protein meal it can be a source of cooking oil and other edible oil products. Tarwi oil is rich in unsaturated acids. It contains high oleic acid and low linoleic acid

Table 3.4 Ecological requirements of grain lupins

Species	*Climate*	*Soil type*	*Nutritive requirements*
L. albus	Cold to moderately warm at vegetation stage. Frost resistant	Sandy, loams; moderately acidic to slightly calcareous; not resistant to water logging but some cvs. may be ± adapted.	Moderate
L. angustifolius	Cold to moderate temperature at vegetative stage. Frost resistant	Sands, loams; moderate to neutral acid content; not resistant to waterlogging	Low to moderate
L. luteus	Moderate temperature at vegetative stage, slightly resistant to frost	Sands, loam; medium to high acid content; slightly tolerant to water logging	Low
L. mutabilis	Cold to moderate some resistance to frost and drought	Sands and loams; moderately acidic	Moderate

Adapted from Dovart (1990)

which gives greater stability to the oil. *L. albus* contains a small amount of erucic acid, an undesirable fatty acid which might cause health hazards if it is present in high amounts.

3.3.1 **Alkaloids**

Seeds and vegetative parts contain many toxic compounds (Table 3.6) which make the crop unsuitable for consumption without treatment. The alkaloids present in seeds are a range of up to 50 quinolizidines which are bitter and toxic. The most important ones are: lupinine ($C_{10}H_{19}NO$),

Table 3.5 Composition of lupin seeds (% dry matter)

Species	*Crude protein (N × 6.25)*	*Average*	*Oil*	*Average*	*Crude fibre*
L. albus	34–45	39	9–15	11	3–10
L. luteus	36–48	42	4–7	5	15–18
L. angustifolius	28–38	34	5–7	6	13–17
L. mutabilis	32–46	43	13–23	18	7–11

Adapted from Dovart (1990), Pate *et al.* (1985)

Table 3.6 Alkaloid contents of lupin seeds

Species	*Alkaloid Range*
L. albus	0.005–0.367
L. angustifolius	0.095–1.414
L. luteus	0.470–1.500
L. mutabilis	0.007–3.970

From: Gross 1990; Luisa and Burao de Costa 1990.

lupanine ($C_{15}H_{24}N_2O$), sparteine ($C_{15}H_{26}N_2$) and hydroxylupanine ($C_{15}H_{24}N_2O_2$) (Table 3.7). Toxicity of a species varies seasonally and geographically and toxicity often increases after flowering.

3.3.2 **Amino Acids**

The proteins of lupins are poor in sulphur-containing amino acids as in other pulses. *L. albus* is also deficient in lysine and the biological quality of

Table 3.7 Alkaloids present in different species of lupin

Alkaloids	*L. albus*	*L. angustifolius*	*L. luteus*	*L. mutabilis*
Isosparteine	X		X	
Sparteine	X	X	X	X
Gramine	X	X	X	X
13-Hydroxylupanine	X	X	X	
Lupininc	X	X	X	X
Angustifoline	X	X	X	X
Cytisine	X	X	X	X
12, 13-Dehydro-angustifoline	X		X	X
Not Identified	X	X		
Not Identified	X			
N-Methylcytisine	X	X	X	X
4-Hydroxylupanine	X	X	X	X
Not Identified	X	X		X
Lamprolobine	X			
Lupanine	X	X		X
X-Isolupanine	X	X		X(?)
17-Oxolupanine			X(?)	X(?)
5-Dehydrolupanine				X(?)
Rhombifoline				X(?)

Source: Lopez-Bellido and Fuentes, 1986

Table 3.8 Amino-acid contents in lupin seeds (g/16 g N_2)

Amino acids	Species and Variety				FAO/WHO reference protein (1973)
	L. albus *Kalina*	*L. angustifolius* *Mirela*	*L. luteus* *Reda*	*L. mutabilis* *Inti*	
Isoleucine	4.4	4.0	3.7	3.9	3.5
Leucine	7.5	7.1	7.9	6.8	4.0
Lysine	4.7	4.6	4.9	5.2	5.5
Methionine	0.7	0.7	0.7	0.8	
Cystine	1.6	1.8	2.4	1.6	
Met. + Cys.	2.3	2.5	3.1	2.4	3.5
Tyrosine	4.6	3.4	2.7	a	
Phenylalanine	3.5	3.7	3.9	a	
Tyr. + Phe.	8.1	7.1	6.6		6.0
Threonine	3.6	3.4	3.2	3.5	4.0
Tryptophan	0.8	0.8	1.0	a	1.0
Valine	3.9	3.5	3.2	3.7	4.0
Histidine	2.2	2.7	2.9	a	

a, present but not determined. Source: Gross, 1990.

protein in *L. albus* and *L. mutabilis* in relation to the protein efficiency ratio (PER) is low but Savage *et al.* (1988) suggested that this can be improved by the addition of methionine. However, the level of essential amino acids of the four grain lupins is comparable to the recommended levels of FAO/WHO (Table 3.8) which suggest that lupins can be used in similar ways to other pulses.

3.3.3 **Other seed chemicals**

Seeds of lupins contain β-sitosterol. Samples of Egyptian *albus* seeds are good sources of Vitamins A, B and C. The genus is also reported to contain dopa, hydrocyanic acid, hypoxanthine, malonic acid, pachycarpine, pectine, piperidine, saponin, sparteine, trigonelline, tryptophan and xanthine.

The trypsin activity in *L. mutabilis* is 1.16 T.I.U. which is considerably less than soyabean (30.1 T.I.U.). A low amount of cyanogenic glucosides is also reported but it is well below the permitted level (Montgomery, 1965).

3.3.4 **Fibre**

The approximate composition of protein and fibres of the various parts of the plant has been given by Kay (1979) as:

1. Stem - crude protein 3.4–11.3%, crude fibre 26.1–54.6%.
2. Leaves - crude protein 9.7–21.5%, crude fibre 21.3–29.9%.

3. Pods - crude protein 3.5–17.3%, crude fibre 33.6–42.2%.

The crude fibre content of the seed coat exceeds 10% in *L. luteus* and *L. angustifolius* but the other two species contain less fibre. This could be reduced further through breeding and is important for palatability. However, the seed coat can be a valuable fibre source for human consumption.

3.4 AGRONOMY, YIELD AND PRODUCTION

Lupins are grown as grain crops or as green manure crops. Their place in rotations can be beneficial. The rotation of lupin with potato and wheat has enhanced the yield and quality (Reeves, 1984). This has been demonstrated in Australia when it is grown as winter-grown grain crops or as 'break crops' in rotation with other crops. *L. angustifolius* has also been rotated in South Africa with cereals and is traditionally a summer grazing grain crop. The grain is either used locally or exported to Europe. Other examples in rotation are *L. angustifolius*, which is grown as a winter forage crop in rotation with summer-grown maize in south eastern USA; *L. mutabilis*, which is rotated with potato, maize and quinoa in the Andes; and in Northern Europe grain lupins are grown as a catch crop for green manure after winter cereals.

The effect of lupin rotation is also felt in breaking disease cycles (Reeves, 1984). However, the influence differs from place to place and on the combination of crops rotated. The benefits of lupins in the rotation for soil improvement and as cash crops has not yet been fully estimated. However, significant increases in yield in a range of crops following grain lupins have been reported in Australia, former USSR, Poland, Germany, France, England, Italy, Kenya, New Zealand, Peru and elsewhere. Reeves (1984) reported that in comparison to other successive cereals, cereal grain yield increases in general from 30 to over 100% when cereals are grown after lupins. Some effects on soil have also been reported. Herridge (1982) clearly indicated the effects of *L. angustifolius* in wheat rotation on sandy soils in Australia following lupins. However, the availability of nitrogen from lupin residues is affected by environmental conditions such as temperature, rainfall, rates of mineralization and losses due to leaching.

3.4.1 **Sowing and general agronomy**

Seed germination is epigeal. Seeds are sown by means of a seed-drill (or by hand by smallholders) at a depth of 1.5–5 cm in a clean seed bed. In warm zones (Mediterranean area, South America, South Africa, Australia and New Zealand) sowing is mostly carried out after the first rain in autumn. In cool-temperate zones (Europe, former USSR, USA) planting is carried out in early spring.

Several investigations have been carried out on spacing. In general a

Table 3.9 Seeding rate of lupins in different countries

Species	*kg/ha*	*Country*	*Comments*
L. albus	100–120	Brazil	
	80–170	Chile	
	160–170	USA	
	70–140	Spain	60–70 cm row distance
	160–180	France	60 plants/m^2
	200	Italy	
	160–180	Portugal	
	120–150	South Africa	
	($0.5–0.8 \times 10^6$)	USSR	45–50 to 15 cm row distance
L. angustifolius	80–100	Australia	
	–	New Zealand	15 cm row distance; 70 plants/m^2
	60–80	South Africa	
	60–100	Bolivia	
	40–50	Ecuador	50–60 cm row distance
	120–150	Peru	
L. luteus	(1.1×10^6)	Poland	
	220–250	Portugal	
	($0.4–0.6 \times 10^6$ to 10^6)	USSR	13–15 cm row distance for forage; 45–50 cm row distance for grain

Source: Lopez-Bellido and Fuentes, 1986.

spacing of 15 cm between plants and 20 cm between rows is used. The seeding rate is: *L. albus*, 70–180 kg/ha; *L. angustifolius*, 40–150 kg/ha; *L. luteus*, 220–250 kg/ha; and *L. mutabilis*, 120–180 kg/ha.

However, seeding rates vary according to varieties used and are different in different countries (Table 3.9). Further studies are needed on the effect of population density and grain yield. *L. angustifolius* germinates quickly but low germination is noted in a wild type such as *L. consentii*. *L. albus*, *L. luteus* and *L. mutabilis* show intermediate germination. However, inoculation with slow growing bacteria 'Phytomyxa' and pre-treatment with sulphuric acid can increase field germination.

Lupins are poor competitors with weeds, and weed control is essential, especially during early stages of growth. The use of pre-emergence herbicides for weed control has been successful (Table 3.10).

3.4.2 **Fertilizers**

It is generally known that lupins do not require high doses of fertilizer as

Table 3.10 Herbicides used in lupin cultivation

Herbicide	*Dosage (kg/ha)*
Pre-sowing	
Triallate	1.4
Benfluraline	1.26
Trifluraline	1.15
Pre-emergence	
Simazine	0.5–1.50
Methabenzthiazuron	2.8
Neburon + nitrofen	2.3 + 1.75
Nitrofen	
Nitrofen + linuron	1.8 + 0.55
Chlortoluron	2.5
Trifluraline + linuron	1.2 + 0.6
Pendimethalin + neburon	0.4 + 1.84
Terbutryn + neburon	1 + 1.5
Metribuzin + linuron	–
Atrazine	1.1
Pronamide	0.8–1.6
Prometryne	-
Alachlor	1.1–2.2
Metribuzin	0.5
Fluometuron	1–2
Terbutryn + tetbuthylazine	–
Propyzamide	–
Diuron	0.4–2.4
Post-emergence	
Diclofop-methyl	0.9
Alloxydine sodium	0.75
Carbetamide	1.5–3
Simazine	0.5
Flamprop-isopropyl	0.6
Propazine	0.75–1.5

Source: Lopez-Bellido and Fuentes, 1986

they can fix nitrogen. However, most authors advise application of nitrogen at the rate of 20–30 kg/ha nitrogen at the beginning of growth. Lupins need high doses of phosphates, usually applied at the rate of 300–500 kg/ha. In South Africa and in the USA, 300–400 kg/ha of superphosphate is applied before sowing or later. In poor soils, such as sandy soils, 28–56 kg/ha potash is also applied. In Kenya superphosphate is applied at the rate of 100 kg/ha and in Australia it is applied at the rate of 373–1119 kg/ha depending on soil but trace elements and potassium and sulphur are also used when required. It is believed that a high rate of phosphate drilled with the seed can affect seedling emergence as lupin seeds are sensitive to concentrations of salts. It has been observed that lupins are sensitive to potassium deficiency and this deficiency can be noticed from leaf senescence and also because susceptibility to *Pleiochaeta setosa* increases because of this nutrient deficiency. The normal application is between 50 and 120 kg of K2 O/ha. Lopez-Bellido and Fuentes (1986) reported that the levels of copper, zinc and molybdenum required for lupins are similar to those for wheat. They also reported that lupins respond well to sulphur fertilizer.

3.4.3 **Harvesting**

Lupins are normally harvested when the pods have turned brown, preferably when the weather is moderately cool. Harvesting is carried out at the beginning of summer for the autumn–winter sowings, and at the end of summer or the beginning of autumn for spring sowings.

A grain harvester with drum speed suitably adjusted to minimize seed injury can be used satisfactorily and the drum speed should be around 350–450 r.p.m. Freshly combined seeds should be cleaned immediately to remove large green seeds and trash. Small lots of seeds can be dried naturally but care should be taken that seeds are stirred at least twice a day. Large lots of seeds can be dried in a commercial drying plant and to a moisture content of about 10% and at temperature below 45°C to minimize the possibility of fungal diseases. The dried seed may be stored in bags or in well-ventilated warehouses. The thick seed coats make it relatively resistant to storage pests.

3.4.4 **Pests and diseases**

Diseases

Several diseases have been reported which attack cultivated lupins (Table 3.11).

The most severe one is brown leaf spot which causes damage to winter grown crops at the pre-flowering period in Australia, France and Brazil. It is transmitted mostly from residues of the previous crop and infects all above-ground parts of the plant causing dark brown to purple necrotic spots. Control methods include chemical treatment of the seed, removal of crop residues by grazing, wider rotations and early spraying.

Another serious disease is caused by anthracnose, particularly in *L.*

Table 3.11 Disease susceptibilities of lupins

Diseases	*L. albus*	*L. angustifolius*	*L. luteus*	*L. mutabilis*
Sowing–emergence				
Pythium sp.	X	X	X	X
Rhizoctonia sp.	X	X	X	X
Thielaviopsis basicola	X	–	X	–
Vegetative period				
Ascochyta caulicola	X	–	–	–
Ascochyta gossypii	–	X	–	–
Ascochyta pisi	–	–	–	X
Botrytis cinerea	X	X	X	X
Colletotrichum gloeosporioides	XX	X	–	XXX
Erysiphe polygoni or *pisi*	X	XXX	X	X
Pleiochaeta setosa				
Winter	XXX	XXX		
Spring	X	X	XX	–
Sclerotina sp.	X	X	X	
Viruses				
Bean yellow mosaic	X	X	XX	X
Cucumber mosaic	X	X	X	X
Flowering, harvest				
Fusarium oxysporum	X	–	XXX	X
Fusarium japonicum	–	XX		–
Fusarium anguioides	X	–	–	–
Oginiosus leptostromiformis	X	XX	X	–
Stemphilium vesicarum	–	XXX	–	–
Uromyces sp.	X	X	X	X

XXX, High susceptibility; XX, medium susceptibility; X, exists in species; –, no information.
Source: Lopez-Bellido and Fuentes, 1986.

mutabilis in the Andean region of South America. This can be controlled by seed disinfection with captane or triurrame and crop spraying.

Wilting and damping-off diseases are also noticeable when the crop is grown at relatively cool temperature and cause damage to lupins in most growing countries. Seed dressing and crop rotations are best control methods.

Grey leaf spot disease causes grey or grey-brown necrotic spots in *L. angustifolius* which is badly affected with this disease in Australia. It is controlled by crop hygiene and restricting cultivation to well-drained soils.

Table 3.12 Lupin species susceptible to pests

Pests	*L. albus*	*L. angustifolius*	*L. luteus*	*L. mutabilis*
Sowing–emergence (soil insects)				
Lepidoptera:				
Agrotis sp.		X		X
Feltia sp.				X
Copitarsia turbata				X
Diptera:				
Phorbia platura	X			X
Tipula sp.	X			
Coleoptera:				
Adioristus sp.				X
Agriotes spp.	X			
Astylus spp.				X
Collembola;				
Sminthurus viridis	X	X	X	
Mollusca:				
Agrioimax agrestis	X			
Acari:				
Halotydeus destructor		X		
Vegetative period				
Lepidoptera:				
Ciampa arietaria		X		
Copitarsia turbata				X
Heliothis punctigera	X	X	X	
Heliothis armigera		X		
Plusia chilensis				X
Diptera:				
Liriomyza sp.				X
Melangromyza sp.				X
Ophiomyia sp.				X
Phytomyza horticola	X			
Coleoptera:				
Epicauta sp.				X
Sitona griseus			X	
Homoptera:				
Acrytosiphon pisum	X		X	

Table 3.12 Continued

Pests	*L. albus*	*L. angustifolius*	*L. luteus*	*L. mutabilis*
Acrytosiphon kondoi		X		
Aphis craccivora	X	X		
Myzus persicae	X	X		
Flowering harvest				
Lepidoptera:				
Heliothis punctigera	X	X	X	
Heliothis armigera		X		
Helicoverpa titicacae				X
Leptotes callangae				X
Peridroma saucia				X
Etuekka behrii		X		
Heteroptera:				
Calocoris norvegicus	X			
Adelphocoris lineolatus	X			
Lygus rugulipenuis	X			
Thysanoptera:				
Frankliniella spp.				X

Source: Lopez-Bellido and Fuentes, 1986.

Bean yellow mosaic virus (BYMV) causes damage to seed stocks of many species but serious damage is mostly confined to sweet lupins. Frey (1982) has found a mycoplasma in Peru which causes abnormal plant growth and 'greening of flowers'.

Another fungal disease, *Phomopsis rossiana*, is of considerable importance since it produces toxins which can cause lupinosis in sheep, cattle and horses when these animals are grazed on mature lupin stubble. The spraying with benomyl before pod formation reduces the amount of fungal infection.

Pests

Table 3.12 lists the different pests which attack lupins. Aphids are important pests of sweet lupins in Australia, Brazil, Spain and New Zealand. *Heliothis* spp. also cause damage to buds, flowers and pods of winter-grown crops in Australia, South Africa, Portugal and southeastern USA. Soil insects, cut worms, mining insects, birds and rodents are other pests which cause damage at the emergence and vegetative stage.

Losses caused by these pests can be partly avoided by planting earlier-maturing cultivars. Biological control methods are also used for limiting the damage caused by *Heliothis* spp. and spraying with DDT, carbaryl, endosulphan, permeth and fenvalerate (Nelson *et al.*, 1984).

Root-knot nematodes are reported to affect lupin crops in the USA and in parts of Australia. All four cultigens are susceptible to nematodes.

In South Africa snails and slugs often present problems and white fringed beetles and grasshoppers have been reported to cause damage to crops in the USA.

3.4.5 **Yields**

It is believed that harvest index, branching structure and inflorescence pattern all influence yield in lupins. This has been demonstrated in *L. albus* (Goldstein, 1984) and in *L. angustifolius* (Bishop *et al.*, 1984). However, the influence of sowing date and plant density has also been recognized (Goldstein, 1984).

The yield of lupins depends on the improvement of yield components. In general, breeders agree that pods/plant, plant density, seed numbers/pod, seed weight, protein and oil contents and compact pod form with long main axis are yield characters. However, there is strong interaction between the yield components and environmental conditions.

Lopez-Bellido and Fuentes (1986) reviewed the situation of yield in lupins and its relation to yield components. Table 3.13 shows the yield/ha of cultivated lupins in different production areas.

Czerwinski and Swiecicki (1990) reported production of yellow lupin in Europe and the world production has been highlighted by Lopez-Bellido and Fuentes (1986). It is believed that lupin production will increase as the EC has set up a special project on alternative products from lupins including protein and oil (Rossetto, 1990).

3.4.6 ***Rhizobium***

Artificial inoculation of lupin seeds with *Rhizobium* is not widely practised except when the crop has not been grown before. However, the strains of *Rhizobium* isolated from lupin nodules are usually slow growing and tolerant of acidic conditions in culture on artificial media.

It is reported (Pate *et al.*, 1985) that nitrogen fixed in *L. angustifolius* starts at five weeks after sowing and continues until reproductive maturity but maximum fixation rates are noticed at mid-vegetative growth. It has been suggested that *L. albus* and *L. angustifolius* fix a small proportion of total

Table 3.13 Indicative yields of lupins

Country	*Species*	*(Yield/ha) kg*
Africa		
South Africa	*L. albus*	1000
	L. albus (irrigation)	3000
	L. angustifolius	750
North America		
United States	*L. albus*	1700–3900
South America		
Bolivia	*L. mutabilis*	1200
Brazil	Without specification	1200–2000
Chile	*L. albus*	2000–5000
	L. luteus	1600–2600
Ecuador	*L. mutabilis*	600–700
Peru	*L. mutabilis*	300–2000
Europe		
Spain	*L. albus*	500–1500
France	*L. albus* (spring)	2500–4000
	L. albus (winter)	3500–5000
Italy	*L. albus*	1480
Portugal	*L. albus*	1200–1500
	L. luteus	800
Poland	*L. albus*	2000–3500
	L. angustifolius	1500–2500
	L. luteus	1200–2000
Oceania		
Australia	*L. angustifolius*	500–2500
New Zealand	*L. angustifolius*	3000–4000
USSR	*L. albus*	2000–4000

Source: Lopez-Bellido and Fuentes, 1986.

nitrogen (about 15%) before flowering, the maximum stage is reached during fruiting and the nodulated root competes efficiently with the developing fruits and other plant parts for the carbohydrates. Lopez-Bellido and Fuentes (1986) state that the total capacity of N_2 fixation by lupin has been estimated to be between 102 and 252 kgN/ha. However, various factors such as waterlogging, level of soil nitrogen or type of fertilizer may influence nodulation and nitrogen fixing.

Young lupin nodules are ovoid to round, firmly attached to the root by a broad base of secondary tissue, and normally abound in leghaemoglobin; ineffective nodules are uncommon. *Rhizobia* enter root hairs in a typical manner and penetrate the root cortex within typical infection threads, but infection strands are not evident in nodule tissue. Host cell divisions account for further dissemination of *rhizobia* (Allen and Allen, 1981).

Nodulation has been widely reported for all four cultigens. Lupin *rhizobia* encompass a wide host range and show parallels in promiscuity with the cowpea *rhizobia* group. The organism involved is *Rhizobium lupini*. Host affinities include *Lupinus* and *Ornithopus*. In culture the growth rate is slow and the cultural reaction is alkaline. There are some affinities to *R. japonicum* of soyabean.

3.5 GENETICS AND BREEDING

3.5.1 Germplasm collections

The germplasm of lupins has hardly yet been exploited. The collection, documentation and evaluation of both Old World and New World species is far from complete. Swiecicki (1988) reported that the collection of the lupins are maintained in six large gene banks and also in fifteen small collections. He believed that there could be about 10000 accessions in these collections and there could be about 3000 duplicates in the collections. However, information on wild species is mostly lacking, particularly the Egyptian and Sudanese ecotypes of white lupins.

An analysis of the current data shows the following institutions maintain substantial collections (AL, *albus*; AN, *angustifolius*; L, *luteus*;, and M, *mutabilis*).

Australia:	West Australia, Department of Agriculture	AN
Bolivia:	Paurumani, Cochabamba	M
Germany:	FAL, Braunschweig/Ziguk Gatersleben	AL,AN, L, M
France:	INRA, Lusignan	AL, M
Netherlands:	CGR, Wageningen	AL, L
Peru:	UNA La Molina	M
	PCA, IICA, Cuzco	M
Poland:	Wiatrowo, Wagroisec	AL, L
Portugal:	Oeiras, EAN	AL, AN, L
Spain:	INIA Madrid	AN, L
	SIA Badajoz	AN, L, M
UK:	Univ. Reading	AL
USA:	NPGS	AL, AN, M

The IBPGR has designated, in agreement with the institutions concerned, genebanks to hold major germplasm collections for long-term security. These are in Germany (for a global collection thereby safeguarding duplicates from elsewhere), Spain (for European material) and a centre has yet to be designated in S. America.

3.5.2 **Gaps in knowledge**

Information on the taxonomy and ecogeographic distribution of lupins is still limited and needs to be studied to determine the diversity of the genus *Lupinus*. Gladstones (1984) and Pate *et al.* (1985) made a review of the genetic resources of the Old World and its genetic variability.

The poor harvest index and yield capacity of *L. luteus* and late maturity in *L. albus* cause problems for improvement. However, *L. angustifolius* seems to have a more diverse genepool of wild and semi-domesticated forms and related species in which there are few barriers. It can be cultivated in a wide range of environmental conditions. It is more drought resistant than *L. albus* and *L. luteus* and grows better in poor soils than *L. albus*. *L. angustifolius* has erect stem and fine leaves allowing good light penetration for yield improvement. However, disease susceptibility of *L. angustifolius* and restricted gene pool in *L. albus* and *L. luteus* inhibit introgression of appropriate germplasm from other species.

Plant breeders have limited accessions evaluated for selection but Pate *et al.* (1985) mentioned that affinities in alkaloids between the Mediterranean species and pollen tube growth and carpellary stimulates in crosses between *L. mutabilis* and European species suggest that gene exchange might be possible between New World and Old World species. In addition the African germplasm is more or less unknown and this has great potential for utilization, particularly for selection of drought resistance.

L. mutabilis has a wide genetic variability illustrating adaptation to microhabitats and natural selection, typical of the diverse Andean environment. Variability has especially been noted in plant form, vegetative growth, susceptibility to frost and disease, protein, oil and alkaloids content. It is believed that the variability in Andean lupin can be utilized to improve Old World lupins.

3.5.3 **Breeding**

Several breeding programmes to improve lupins have been initiated in Australia, Europe and in Latin America with the following objectives: a) higher yields and quality, b) pests and disease resistances, c) zero alkaloids contents, d) high protein and oil contents, e) balanced protein quality, f) drought and frost resistance, g) tolerance to lime and salinity, h) non-shattering types, i) cultivars not requiring vernalization. In addition bitter type lupins may be important for by-products and non-food uses (such as Lupinex, see Section 3.6).

The breeding for low alkaloids in *L. albus*, *L. luteus* and *L. angustifolius* has been in progress for a long time and low and acceptable mutants have been obtained. Blanco and Jimenez (1982) have selected lines of *L. mutabilis* with low alkaloid (0.1%) which have a polygenic character. Williams *et al.*

(1984) also reported an allele-recessive mutant of low alkaloid content (0.2–0.3%). Several authors have reported increases in protein and oil content and found that both characters are negatively correlated in *L. mutabilis* and *L. ablus* but selection of large seeded types could solve this problem.

The breeding for early maturing and frost resistant cultivars of *L. albus* has been in progress in most growing regions. The selection of lines with pod setting on the main stem and compact form may increase seed yield (Mikolajczyk *et al.*, 1984). Mutants of non-shattering pods and permeable seed coats were targeted in Germany. Gladstones (1984) reported the development of lines of *L. luteus* combining these characters and disease resistance especially to *Fusarium* spp. In Europe and in the former USSR efforts have been made to select lines with little branching which are early maturing with high pod setting and mutants have been found for pod setting types with few branches which are now used to seek early maturing cultivars. Swedish scientists selected the sweet and permeable seed coat lines of *L. angustifolius* in the mid-1940s and Gladstones reported commercial cultivars of Unicrop with combined characters of early maturing, white flowers and non-shattering pods in the 1970s.

Lopez-Bellido and Fuentes (1986) reviewed the breeding work on *L. angustifolius* and discussed the primary objectives of *L. angustifolius* development. Several studies (Lopez-Bellido and Fuentes, 1986; Fuentes, 1985; Reeves *et al.*, 1977; Perry and Poole, 1975) have discussed growth patterns, influence of vernalization, branching structure and floral initiation. They have also reported selection objectives for physiological growth as lupins are grown in diverse climatic regions. These include a high rate of photosynthate production, better economy of photosynthate use in vegetative growth and partitioning profiles to favour fruit development. Higher plant density and determinate growth habit with little branching can increase the dry matter. Higher plant density is also correlated with higher yield but this needs further investigation as environment plays a role for this parameter (Chaudhary and Haq, unpublished). Furthermore, the efficiency of utilization of net photosynthate by roots and shoots do not differ significantly between genotypes and might not be amenable to species improvement. The conversion of photosynthate and nitrogen fixation into dry matter and protein in seeds differs between landraces and cultivars. This is due to less carbon and nitrogen assimilation in non-harvestable products in developed cultivars of determinate habits. The other important factor is the harvest index which relates to photosynthate products and consequently the nitrogen for both seed yield and protein. The accumulation and utilization of dry matter and its relation to fruiting, vegetative parts development and modification of lupin growth patterns using growth regulators have been discussed by Lopez-Bellido and Fuentes (1986).

3.6 FUTURE POTENTIAL

Lupins are not yet important in international trade but export from Australia has been reported to have exceeded 100 000 t (Dovart, 1990). South Africa and Europe primarily import from Australia. Recently, the USA has also started to import lupins.

Rossetto (1990) and others believe that the lupin product can be used as a substitute for soyameal and its market price would have to be competitive. He also suggested that the product Lupinex can make its position in international trade soon. Lupinex is a bitter compound extracted from seeds. It has been suggested that Lupinex can be used to protect plants from insects and diseases (Kahnt and Hijazi, 1990). There are a number of minor uses: Lupin seed has been used as a substitute for coffee and its flour is reported to have been used as an egg substitute in Hungary. Also seeds of *L. angustifolius* and *L. albus* can be used as an excellent source of asparagine, used for commercial production of tuberculin. Yellow lupin, *L. luteus*, is also a good honey plant and is widely introduced as a fragrant ornamental. None of these uses are likely to be expanded widely.

The major potential will be a role in soil rehabilitation and afforestation associated with the rural poor who can also use the products.

L. albus has shown improvement in soil structure when tested in Brazil. *L. mutabilis* has been used to stop soil erosion of highlands in Peru. *L. arboreus* has also been used in New Zealand to stabilize sand dunes in coastal areas. The use of *L. mutabilis* in agroforestry with *Pinus* and *Eucalyptus* in Peru, *L. arboreus* with *Pinus* in New Zealand and *L. polyphyllus* with *Picea* spp. in Germany has already been reported by Gross (1982), Mead and Gadgil (1978) and Melzer and Luck (1984).

Further application in areas of high population will show the value of lupins in major crop diversification programmes (especially on irrigated areas and dry areas after the major staple crops). Agroforestry will be pressingly important. Lupins clearly have a role in soil rehabilitation.

Other research, especially in the developed world, will relate to value-added products. The EEC is mostly interested in non-food and high protein and oil content species of lupins for European countries. In this matter, the potential of *L. mutabilis* should be explored. The selection of compact type with long axis will be useful to grow in Europe, particularly in the UK where *L. albus* produces low yields due to late maturity. The genetic variability of *L. mutabilis* has not yet been explored and wider efforts should be made in this direction. Lastly, wide hybridization will be a useful tool for transfer of important agronomic traits among species.

3.6.1 **Lupins in human nutrition**

Grain lupins have been used for human consumption for thousands of years. However, their utilization in *modern* food production is still limited. Pompei (1984) reported that this utilization is complex as the products (flours, oils, concentrates and isolates) need to meet quality and nutritional requirements. The nutritional protein value of lupin is significantly increased when supplemented by synthetic methionine at a level of 2% of total protein (Lopez-de-Romana, 1982). Investigations have shown that lupin is a low cost protein source and is best used in combination with other foods.

Several authors (Reyes *et al.*, 1980, Reyes, 1982) have used 10–20% lupin flour with wheat flour in making bread. It produced a good sized and good textured bread and had pleasant colour with high protein and improved amino acid and it was good from the consumer's point of view. Lupin is also used in biscuit and cake making, pastas, lactic substitutes and is incorporated in the production of baby foods, snack foods, hamburgers, etc. In Peru, lupin flour is added to biscuits, bread, noodles, sauces and soups. Powdered milk is also produced from lupins by spray drying at 17°C and is found to be nutritive and acceptable. Tincan packaging has also extended the shelflife to a minimum period of six months at normal conditions. Camacho *et al.*, (1988) reported that a powdered milk plant in Chile is technically and economically feasible and can contribute to increasing the income of farmers of that region.

The oil of *L. mutabilis* has good nutritive qualities comparable to other vegetable oils and it is more digestible than mixtures of soya and cotton seed oil.

Gross (1990) reported that schoolchildren did not show any reactions when they consumed different lupin products along with other foodstuffs during allergy tests in Peruvian schools over a period of three years. He has also suggested several methods of debittering and using lupins. He believes that the wider introduction of lupin products for human food is not only a matter of research in food chemistry and technology but also a challenge to food and agricultural policy.

3.6.2 **Lupins in animal nutrition**

Lupins have been used for animal feed for many centuries. Lupin seed is used as a source of protein in concentrate mixtures for feed. In Australia it is used increasingly as a substitute for groundnut cake, fish meal, or soyameal in the production of high quality livestock rations. Lupin flour is being used in the manufacture of dry dog biscuits in Australia and research has indicated that it could also be used as a protein supplement in some other types of pet foods.

Although experiments were carried out for nutritional feed value as early as 1930 the role of carbohydrates, the presence of antinutritive compounds or availability of vitamins are not fully clear. Guillaume (1980) reviewed current uses in animal nutrition. Lupins are mixed with other fodders to compensate the deficiency of sulphur amino acids. Lupin is normally used for protein in the fodder ration at a level of 10–20% of the total diet depending on the type of animal. This varies for ruminants or monogastrics, for which the main problems are still methionine deficiency and the high quantity of testas which are rich in fibre. Lupin may be used for ruminants as green forage, in grain, or as standing stubbles for sheep. The situation is more complex in monogastrics: lupin is limited by methionine deficiency and the low nutritional quality of its carbohydrates.

The carbohydrates are practically indigestible to birds for unknown reasons. Several experiments on digestibility have been carried out on weaning piglets, swine, young calves, milking cows, lambs, broiler and egg-laying hens and even fish (carp and trout) to observe its advantages and disadvantages when used to replace maize and soyabean (Herrera *et al.*, 1980; Emile *et al.*, 1988) and a respectable information base is developing.

Figure 3.1 A lupin crop.

REFERENCES

Allen, O.N. and Allen, E.K. (1981) *The Leguminosae. A Source Book of Characteristics, Uses and Nodulation*, University of Wisconsin Press, Madison, Wisconsin.

Bishop, A., Hamblin, J. and Nelson, P. (1984) Yield and harvest index of branches and reduced branching Lupins (*Lupinus angustifolius*). Proceedings of 3rd International Lupin Congress, pp. 586–7.

Blanco, O. and Jimenez, S. (1982) Avances en la reducción genetica del centenido de alcaloides del tamui (*Lupinus mutabilis* Sweet) Advances in the genetics reduction of alkaloid content in tamui (*L. mutabilis*), in *Actas/II conferencia Internacional del Lupino* (Torremolinos, España - Mâyo 1982), pp. 46–8 (ed. Luis Lopez Bellido) Servicio de Publicaciones Agranas, Ministerio de Agricultura, Pesca y Alimentacion, Madrid.

Camacho, L., Molina, R. and Vargas, E. (1988) Economic engineering studies for the installment of a lupin milk processing industry. *Proceedings of 6th International Lupin Conference*, (ed. T. Twardowski), PWRiL, Ponzan, Poland, pp. 672–81.

Czerwinski, T. and Swiecicki, W. (1990) Lupin-breeding aspects in a semi-continental European climate, in *Lupin Production and Bio-processing for Feed, Food and other By-products* (eds Y. Birk, A. Dovart, M. Waldman and C. Uzurean), pp. 57–67.

Darlington, C.D. and Wylie, A.P. (1955) *Chromosome Atlas of Flowering Plants*, George Allen and Unwin, London.

Dovart, A. (1990) Aspects of grain lupin cultivation. An overview, in *Lupin Production and Bio-processing for Feed, Food and Other By-products* (eds Y. Birk, A. Dovart, M. Waldman and C. Uzurean), pp. 5–8.

Duke, J. (1981) *Handbook of Legumes of World Economic Importance*, Plenum, New York and London.

Dunn, B.D. (1984) Cytotaxonomy and distribution of New World lupin species. *Proceedings of 3rd International Lupin Conference*, pp. 67–86.

Emile, J.C., Huguet, L., Hoden, A. *et al.* (1988) Sweet Lupin seeds for dairy cows and young bulls' feeding. *Proceedings of 6th International Lupin Conference*, (ed. T. Twardowski), PWRiL, Ponzan, Poland, pp. 383–95.

Frey, F. (1982) Problemas fitosanitarios de los lupinos cultivados en los Andes sudamericanos (Pests and diseases of the lupin crop in the Andes) in *Actas/II Conferencia Internacional du Lupino* (Torremolinos, España - Mayo 1982), pp. 150–3 (ed Luis Lopez Bellido). Servicio de Publicaciones Agrarias, Ministerio de Agricultura, Pesca y Alimentacion, Madrid.

Gladstones, J.S. (1974) *Tech. Bull. 26*, Dept. of Agriculture, W. Australia.

Gladstones, J.S. (1984) Present situation and potential of Mediterranean/African lupins for crop production. *Proceedings of 3rd International Lupin Congress*, pp. 17–37.

Goldstein, W. (1984) Growth types of the white lupin and their suitability for cropping in the Pacific Northwest USA. *Proceedings of 3rd International Lupin Congress*, pp. 558–559.

Gross, R. (1982) Some consideration about future .investigation on lupin. *Lupin Newsletter*, **4**, 27–8.

Gross, R. (1990) Nutrients and anti-nutritive substances of lupin for human consumption, in *Lupin Production and Bio-processing for Feed, Food and Other By-products* (eds Y. Birk, A. Dovart, M. Waldman and C. Uzurean), pp. 164–76.

Guillaume, J. (1980) Nutritive value of lupines seed in farm animals. pp 665–692, in *Agricultural and Nutritional aspects of Lupines: Proceedings of the First International Lupine Workshop, Lima-Cuzco, Peru, 12–21 April, 1980* (eds. R. Gross and E.S. Bunting).

Herera, H., Nuñez, H., Molina, B., Mermoud, J. and Pena, R. (1980) Feeding of rainbow trouts using lupine. pp 653–663. in *Agricultural and Nutritional aspects of lupines: Proceedings of the First International Lupine Workshop, Lima-Cuzco, Peru, 12–21 April, 1980* (eds. R. Gross and E.S. Bunting). Eschborn: Deutsche Gessellschaft fur Technische Zusammenarbit.

Herridge, D.F. (1982) Assessment of nitrogen fixation, in *Nitrogen Fixation in Legumes*, (ed. J.M. Vincent), Academic Press, N.Y., pp. 123–6.

Kahnt, G. and Hijazi, L.A. (1990) Effect of bitter lupin extract (Lupinex) R on yield of grain crops and vegetable plants, in *Lupin Production and Bio-processing for Feed, Food and Other By-products* (eds Y. Birk, A. Dovart, M. Waldman and C. Uzurean), pp. 184–7.

Kay, D.E. (1979) *Lupin (Lupinus) as Food Legumes*, Crop and Product Digest No.3, Tropical Products Institute, London, pp. 246–65.

Lopez-Bellido, L. and Fuentes, M.C. (1986) Lupin crop as an alternative source of protein. *Adv. Agro.*, **40**, 239–95.

Lopez de Romano, G. (1982) Human nutrition, pp. 406, in *Actas/II Conferencia Internacional del Lupino* (Torremolinos, España – Mayo 1982) (ed Luis Lopez Bellido), Servicio de Publicaciones Agrarias, Ministerio de Agricultura, Pesca y Alimentacion, Madrid.

Luisa, M.D. and Burao de Costa, M. (1990) Aspects of lupin composition as food, in *Lupin Production and Bio-processing for Feed, Food and Other By-Products* (eds Y. Birk, A. Dovart, M. Waldman, and C. Uzurean), pp. 94–106.

Maire, R. (1987) *Flore de l'Afrique du Nord*, Vol. 16, Editions Lechevalier, Paris.

Mead, D.J. and Gadgil, R.L. (1978) Fertilizer use in established radiator pine stands. *New Zealand Journal of Forestry Science*, **28**, 105–34.

Melzer, E.W. and Lucke, E. (1984) The effect of *Lupinus polyphyllus* (Lindl.) on *Picea abies* (L.) Karst. in forest ecosystem. *Proceedings of 3rd International Lupin Congress*, pp. 228–36.

Mikolajczyk, J., Stawinski, S. and Wiza, M. (1984) Directions actuelles de l'amelioration et l'etat actuel des Recherches sur l'acclimatisation du Lupin Blanc en Pologne. *Proceedings of 3rd International Lupin Congress*, pp. 570–1.

Montgomery, R.D. (1965) The medical significance of cyanogen in plant foodstuffs. *Amer. J. Clin. Nutr.*, **17**, 103–13.

National Academy of Sciences (1979) *Tropical Legumes: Resources for the Future*. NAS, Washington, D.C.

Nelson, P., Ralph, C. and Hamblin, J. (1984) Development of Lupin Industry in the Geraldton regions. A case study in extension. *Proceedings of 3rd International Lupin Congress*, pp. 590–1.

Pate, J.S., Williams, W. and Farrington, P. (1985) Lupin (*Lupinus*) spp., in *Grain Legume Crops* (eds R. Summerfield and E.H. Roberts), Collins, London, pp. 699–746.

Perry, M.W. and Poole, M.L. (1975) Field environment studies on Lupin 1. Developmental pattern in *Lupinus angustifolius* L., the effects of cultivar, site and planting time. *Aust. J. Agr. Res.*, **26**, 81–91.

Planchaelo, A.M. (1985). Taxonomic Studies of *Lupinus* in South America, in *Actes du IIIe Congres International du Lupin, La Rochelle (France), 4–8 Juin 1984*, pp. 39–54 (L'Association internationale du Lupin).

Plitman, V., and Heyn, C.C. (1984) Old world *Lupinus*: taxonomy, evolutionary relationship and links with New World Species. *Proceeding of 3rd International Lupin Conference*, pp. 55–66.

Pompei, C. (1984) The technologies for lupin valorization in human and animal nutrition. *Proceedings of 3rd International Lupin Congress*, pp. 397–420.

Reeves, T.G. (1984) Lupins in crop rotation. *Proceedings of 3rd International Lupin Congress*, pp. 207–26.

Reeves, T.G., Boundy, K.A. and Brook, H.D. (1977) Phenological development study with *Lupinus angustifolius* and *Lupinus albus* in Victoria. *Austr. J. Exp. Agric. Anim. Husb.*, **17**, 637–44.

Reyes, M., Gross, U. (1982) Utilizacion del *Lupinus mutabilis* en la alimentacion humane. In, *Actas/II Conferencia Internacional del Lupino* (Torremolinos, España – Mayo (1982) (ed. Luis Lopez Bellido) Servicio de Publicaciones Agrarias, Ministerio de Agricultura

Reyes, J., Gross, U., Pounce, L. and Sherpa, M. (1980) Experiments on the use of *Lupinus mutabilis* for human consumption pp. 707–20, in *Agricultural and Nutritional aspects of lupines: Proceedings of the First International Lupine Workshop, Lima-Cuzco, Peru, 12–21 April 1980* (eds R. Gross and E.S. Bunting). Eschborn: Deutsche Gessellschaft fur Technische Zusammenarbit.

Rossetto, G. (1990) Lupin production and processing study, in *Lupin Production and Bio-processing for Feed, Food and other By-products*, (eds Y. Birk, A. Dovart, M. Waldman and C. Uzurean), pp. 70–5.

Rumbaugh, M.D. (1990) Special purpose forage legumes, in *Advances in New Crops*, (eds J. Janick and J.E. Simon), Timber Press, Portland, Oregon, pp. 183–90.

Savage, G.P., Hill, G.D and Bala, A. (1988) Variation in oil, nitrogen and methionine level in *Lupinus mutabilis. Proceedings 6th International Lupin Conference*, (ed. T.Twardowski), PWRiL, Ponzan, Poland, pp. 577–81.

Simpson, M.J.A . (1986a) Geographical variation in *Lupinus albus* L. I Iberia. *Zeit. F. Pflanzenzuchtung*, **96**, 232–40.

Simpson, M.J.A. (1986b) Geographical variation in *Lupinus albus* L. II North-west Spain, the Nile Valley, the Balkans and Turkey. *Zeit. F. Pflanzenzuchtung*, **96**, 241–51.

Smartt, J. (1990) *Grain Legumes*, Cambridge University Press, N. Y.

Swiecicki, W. (1988) Lupin gene resources in the Old World. *Proceedings of 5th International Lupin Conference*, (ed. T. Twardowski), PWRiL, Ponzan, Poland, pp. 2–14.

Thulin, M. (1983) Leguminosae of Ethiopia. *Opera Botanica*, **68**, 1–223.

von Bauer, E. and Gross, R. (1977) Auslease bittersloffarmer formes von *Lupinus mutabilis. Zeit. F. Pflanzenzuchtung*, **79**, 52–8.

von Bauer, E. and Gross, R. (1978) Selection of low-alkaloid forms of *Lupinus mutabilis* as an alternative to the soya bean. *Plant Research and Development*, **7**, 60–6.

Whyte, R.O., Nilsson-Leissner, G. and Trimble, H.C. (1953) *Legumes in Agriculture*, FAO, Rome.

Williams, S.W. (1979) Studies on the development of lupins for oil and protein. *Euphytica*, **28**, 481–8.

Williams, W., Harrison, J.E.M. and Jayasetcara, S. (1984) Genetic control of alkaloid production in *Lupinus mutabilis* and the effect of a mutant allele mutal isolated following chemical mutagenesis. *Euphytica*, **33**, 811–17.

Zohary, D. and Hopf, M. (1988) *Domestication of Plants in the Old World*, Clarendon Press, Oxford.

4

Lagos spinach (*Celosia* sp.)

T. Badra

4.1 HISTORY AND USES

Leafy vegetables are an important part of the daily diet and provide essential vitamins and minerals. Human cultures have often come to regard many such vegetables as spinach-like due to the early spread of *Spinacia oleracea* from Southwest Asia and its use by boiling as a 'pot-herb'. Each area of the world developed its own indigenous group of leafy vegetables: some have become almost ubiquitous, such as cabbage; others have remained more or less localized, many of which have remained underutilized.

Celosia, quailgrass, Lagos spinach, or Shokoyokoto (*Celosia argentea* L.) is a popular leafy vegetable in the Old World tropics (Figure 4.1). It is a member of the Amaranthaceae, a family closely related to Chenopodiaceae and these two families have provided a number of cultivated plants: amaranth, *Celosia*, cockscomb, *Atriplex*, *Chenopodium* and beet, nearly all used as vegetables and/or seeds. Except for beet they are all underutilized and probably have remained in use due to a tendency to be weedy and to show plasticity; often use has related to poorer segments of the population.

C. argentea is the most important species of the genus. It is of African origin although another cultivated form, var. *cristata* Kuntze, was probably domesticated in Asia. *Celosia* has become dispersed throughout the tropics. In its centre of origin, Senegal to Cameroon in Africa, it is both cultivated and wild, found as a weed or ruderal in open waste places of the forest zone. It is an annual herb with wide morphological variation attaining up to 2 m in height with a characteristic flower spike, elongated and pointed and usually silvery-white but often coloured pink to red. This characteristic provided the generic name *Celosia* from the Greek kelos meaning 'burning'. The plant flowers throughout the year in West Africa and leaves (and stems) provide a ready source of vegetables or pot herbs. It has been

Pulses and Vegetables. Edited by J. T. Williams. Published by Chapman & Hall, 2–6 Boundary Row, London SE1 8HN. ISBN 0 412 46610 4.

Figure 4.1 *Celosia argentea* L., leaf and spike.

introduced widely around the tropics and is cultivated particularly in parts of C. America, Martinique and Guadeloupe, the areas adjacent to the northern coast of S. America, and in S., S.E. and E. Asia.

A review of tropical vegetables carried out for the International Board for Plant Genetic Resources (IBPGR) led to leafy amaranths becoming an international priority (Grubben, 1977) although it was recommended that *Celosia* be collected and investigated since it is important for home garden production and marketing. Amaranth has received continuing research interest.

Interest in wider use of *Celosia* in Asia has been expressed (Pal and Khoshoo, 1983). *Celosia* also became the subject of research in a development project located at the Nigerian Horticultural Research Institute funded by UNDP. This was near areas where *Celosia* is part of the local culture. The Yorubas of S.W. Nigeria and Benin associate the vegetable with a big belly and include the plant in incantations. The name shokoyokoto means 'make husband fat' and a local name ajefous means 'good to eat'. Under this name, use of the plant travelled to Sierra Leone with change of name to shokotoyokoto. In West and Central Africa, apart from use as a vegetable, leaves can be used to flavour rice and for feeding animals. In Zaire and in Senegal a fibre is prepared from the stems (Berhaut, 1971).

The leaves are without medicinal use in West Africa but *Celosia* became widely integrated into folk practices elsewhere. Leaves are used in poultices in China for sores and wounds and in India for inflamed areas and abscesses, as well as an antidote for snake bite and the root as a specific for colic, gonorrhoea and eczema. In East Africa the water in which leaves and other parts of the plant have been boiled is used as a baby wash for convalescents by the Masai and in Ethiopia the seeds are used for diarrhoea and the flowers for dysentery and muscular troubles. In S.E. Asia the flowers are considered medicinal for conditions whose symptoms include discharge of blood, e.g. dysentery, and menstruation. The seeds contain an oil. They are used for diarrhoea in Zaire, and in Asia they are considered antiscorbutic and antihelminthic. Seeds are used more widely for a variety of medicinal uses in West Africa (Dalziel, 1937).

In parts of Zaire the plant enters into a form of witchcraft to ensure good crops where it is growing. An interesting parallel to this is recorded from New Britain, Papua New Guinea, where it is planted in taro plots in the belief that the taro smells it and hence produces large tubers. In Central Indonesia it enters into most ceremonies to propitiate protective spirits.

Although *C. argentea* is the major species used there are a number of related minor species:

1. *C. trigyna* L., a weed of cultivation and waste places in the savanna and forest zones throughout West Africa and widespread in East and Central Africa, Madagascar and Arabia. The leaves of this species are bitter but they are widely used as a famine food. It is an incipient domesticate and sometimes seeds can be consciously sown. Leaves are used in soups, sauces and seasonings and are occasionally seen in markets in West

Africa. Excess consumption is reported to cause vertigo or delirium. All livestock are reported grazing it in Senegal though in Sudan there are opposing statements on its acceptability.

This species is widely used medicinally with such diverse applications that it merits wider screening. It is rich in saponins and some flavone. Further information can be found in Burkill (1985).

2. *C. globosa* Shinz., a straggling herb of forest undergrowth, clearings and as a weed. Distribution spans S. Nigeria, W. Cameroon to Zaire and Uganda. The leaves are eaten as a vegetable in Zaire.
3. *C. isertii* Townsend, syn. *C. laxa* Schum. and Thonn., a straggling herb becoming sub-lianeous, found in damp sites in clearings, stream banks and secondary forest. Distribution spreads from Senegal through Central Africa to Tanzania, Zambia and Angola. The leaves are often eaten in West Africa, Gabon and Zaire.
4. *C. leptostachya* Benth., a straggling annual of undergrowth and roadsides distributed from Sierra Leone to Zaire. The leaves are eaten as a vegetable in Zaire.
5. *C. pseudovirgata* Schinz., syn. *C. bonnivairii* Schinz., an upright herb of forest, roadsides and as a weed. It is recorded in S. Nigeria and W. Cameroon but also occurs in Zaire. The leaves are eaten as a vegetable in Zaire.

One other species, *C. cristata* L., is closely related to *C. argentea* of which it was once thought to be a variety. It is widely known as an ornamental plant, the cockscomb, since strains have fasciculate brightly coloured inflorescences or loose feathery coloured inflorescences. Many cultivars have been developed and this species is probably the most documented of all *Celosia* sp.

There are probably over 50 other lesser-known species of the genus (Allison, 1989; Harvard University, 1968; Taylor, 1959; Townsend, 1973, 1982) from Mexico, Peru and Ecuador; Africa; and the Indian subcontinent and elsewhere. Virtually nothing is known about genetic relationships of the wild species nor the genepool limits of the cultigen, although the classical-type taxonomy documents basic morphological data (Appendix).

With wider interest in sustainable cropping systems, especially in the tropics, *Celosia* could have a much more important role especially if seed selections and cultivars were more widely available. With increasing population in many tropical areas, home gardens with their traditional wide diversity will be important continuing sources of vitamins and minerals. *Celosia* tolerates shade somewhat better than *Amaranthus* and evidence points to both complementing each other in terms of growth conditions and disease and pest susceptibility (Grubben, 1975). Seed production with such annual crops does not present major constraints. There could be advantages to maintaining both species in rotation in smallholder situations in order to avoid monocropping with its attendant problems. More importantly there are significant market demands in tropical areas that are not satisfied by adequate production of these leafy vegetables.

4.2 TAXONOMY AND ECOLOGY

The genus is placed in the tribe Celoseae of the subfamily Amaranthoideae. The genus *Celosia* is characterized by stamens being tetrasporangiate, becoming bilocular at anthesis with two lines of dehiscence, plants with alternate leaves, ovary multiovulate, and tepals being scarious (Eliasson, 1988). (The genus *Amaranthus* is distinguished from *Celosia* by its uniovulate ovary.)

Species of interest as vegetables are annual herbs with simple, usually entire leaves. Flowers are bisexual; stamens are united at the base to varying degrees to form a cup, stigmas are filiform and bilobed. There is considerable variation in the number of ovules. The flowers are small, white, yellowish or brown, in bracteate interrupted spikes. The perianth has five equal, spreading, glabrous tepals. Seeds are black, usually strongly compressed and shiny, roundish, feebly to strongly reticulate, grooved or tuberculate. The 1000 seed fresh weight is about 1 g.

C. argentea differs from other West African species in the perianth segments being 6–10 mm long and linear lanceolate and the spikes are single, continuous and very dense flowered. Also the leaves are linear to linear-lanceolate. *C. trigyna* has smaller perianth segments, 2–5 mm long, which are oblong to oblong-elliptic and the spikes are interrupted with short lateral clusters of flowers. *C. globosa* has globose inflorescences, *C. leptostachya* has very long inflorescences with widely-spaced flower clusters, *C. laxa* is somewhat similar to *C. trigyna* but seeds are much more reticulate and not glossy and *C. pseudovirgata* is similar to *C. trigyna* and *C. laxa* but it is an upright herb whereas *C. laxa* is straggling. Keys and other details are to be found in Hutchinson *et al.* (1966).

The morphological variation in *C. argentea* is shown in plant height, leaf size and inflorescence size as well as colouring. There is also the tendency for populations to be heterogeneous due to their need for pollination by insects but with a high degree of self-fertilization. This along with spatial isolation maintains particular local cultivars.

The continuum of wild forms, weedy types and cultivated forms show the following ranges: height 0.4 to 2 m; leaf shape lanceolate-oblong to narrowly linear, acute to obtuse and leaf size 2–15 cm × 0.1–3.2 cm; inflorescences dense spikes 2.5–20 cm × 1.5–2.2 cm. The spikes may have varying degrees of anthocyanin colour thus ranging from green through pink to red.

Local selection has related to particular harvesting practices: plants are either allowed to grow and are then pulled up and the leaves stripped or plants are cut periodically and regrowth takes place; and also to taste, especially degree of bitterness of cooked leaves.

The ecology is broad since *Celosia* can be found in disturbed ground both in cultivated fields and waste ground. In the drier regions it can form flushes in dried-up river beds. On the whole the distribution is patchy, often being locally abundant and then not seen for some distance. *C. trigyna* is also weedy but more associated with savannah and forest regions.

Tindall (1983) outlined the environmental response of a 'green' and a 'red' form of cultivated *C. argentea* as tolerant to a wide range of soil conditions, although a high level of organic material in the soil is required for the production of optimum yields, particularly for the green form. A temperature range of 25–30°C is suitable; both forms appear to respond to fairly stable high temperatures. High levels of rainfall are unlikely to limit growth but irrigation may be necessary during particularly dry periods, especially for the red form which is more sensitive to drought conditions. Both forms grow well at low altitudes, where temperature fluctuations are limited.

Daylength variation does not appear to affect this crop to any significant extent although relatively high light intensities are required to maintain regular leaf formation and development. Since they are short-day plants, flowers are rarely formed under long days (see Section 4.3).

4.2.1 *Celosia* in weedy situations

C. argentea, as a desirable cultivated vegetable, can have its normal growth and development in the field affected by some problematic weed species if the latter are left unchecked. *Celosia* and these species are 'ruderals' or 'pioneers' and it is to be expected that cultivation in the tropics will result in the presence of noxious weeds such as *Imperata cylindrica, Panicum maximum, P. repens, Cyperus* sp., *Brachiaria* sp. and *Amaranthus* sp. (Usoroh, 1989). Another troublesome biomass of weeds associated with vegetable *Celosia* cultivated in the tropics comprised numerous narrow- and broad-leaved weeds, to name a few, *Amaranthus spinosus, Andropogon gayanus, Axonopus compressus, Brachiaria mutica, Cenchrus ciliaris, Chloris gayana, Convolvulus arvensis, Cynodon dactylon, Digitaria eriantha, Sida* spp., *Synedrella nodiflora, Talinum triangulare* and *Tripsacum laxicum.*

Conversely, *Celosia* can be weedy in other farming systems of banana, plantain, pineapple, citrus, mango, guava, avocado and papaya plantations (Usoroh, 1989), and fields grown to tomato, onion, capsicum pepper, okra, groundnut and intercropped sorghum – pigeonpea, green gram and soybean (Kumar and Singh, 1985; Marnotte, 1984; Rajan *et al.*, 1984; Rao *et al.*, 1982; Shelke *et al.*, 1985; Singh *et al.*, 1985). *Ageratum conyzoides, C. trigyna, Commelina benghalensis* and *Euphorbia heterophylla* weed populations in a humid tropical climate in Cote d'Ivoire were found to vary most markedly with changing seasons and different levels of the fertilizer applied to the cultivated crop (Marnotte, 1984). In Pantnagar (India) soybean seed yields were decreased by 80.9% due to an uncontrolled weed community predominated by *Echinochloa colonum, Dactyloctenium aegyptium, Cyperus rotundus, Digitaria sanguinalis, Trianthema monogyna* and *Celosia argentea* (Singh *et al.*, (1985)). The ecological balance of weed species present in irrigated groundnuts was studied (Rajan *et al.*, 1984) using three herbicidal treatments and a hand-weeding treatment at different intervals. *C. argentea* was prominent.

C. argentea shows strong competitive advantages in soils which tend to be clay soils and relatively fertile (Okusanya and Ahmedu, 1983).

4.3 PHYSIOLOGY OF GROWTH AND DEVELOPMENT

4.3.1 **Growth analysis**

Amaranthus cruentus and *C. argentea* are both quick-growing vegetables. The former differs from the latter in having faster initial development, higher dry matter content, shorter generative phase and higher seed production (Grubben, 1977).

Preliminary experiments were conducted by Grubben to analyse growth under two harvesting methods, (1) by uprooting plants and (2) by regular cuttings (Table 4.1).

Table 4.1 Leaf production over time in weeks after transplanting

Weeks after transplanting	*Production of fresh weight g m^{-2}*	*Production of dry weight g m^{-2}*
4	1206	122
6	1578	160
8	1817	185
10	2040	205

Table 4.2 The effect of PAR on leaf weights, and leaf/stem ratios at 28 days from emergence

Species	*PAR* (μmol photons $m^{-2}s^{-1}$)	*Leaf* Fresh wt (g plant^{-1})	*Leaf* Dry wt (g plant^{-1})	*Leaf/stem ratios* (Dry wt. basis)
Amaranthus	215	24.76 ± 2.77	2.30 ± 0.29	3.04
Green Head	325	34.11 ± 2.64	3.46 ± 0.41	2.43
	400	52.27 ± 4.94	6.21 ± 0.55	1.99
Celosia	215	28.10 ± 2.67	2.34 ± 0.23	4.01
IITA (SRV8)	325	43.38 ± 2.45	3.86 ± 0.10	4.34
	400	59.10 ± 2.09	5.19 ± 0.20	3.94

Means ± SE of two experiments each consisting of three plants per treatment.

The total edible product of the harvest, after harvesting by uprooting, includes about 93% of the plant parts after six weeks and about 74% after 10 weeks. Net assimilation rate is less for *Celosia* than for *Amaranthus* but leaf area index is higher in *Celosia*, especially when plants are not cut periodically.

Grubben's results provided overall information on growth. Harvesting by uprooting yields the highest leaf production in about 70 days from transplanting just before plants become reproductive. Local producers tend to harvest earlier than 70 days because quality is better (more tender leaves and less stem material). Good cultivation can increase leaf production to as high as 5, 10 or more t/ha.

Leaf production is related to temperature regime, according to Grubben's experiments. *Celosia* shows no increase of the photosynthesis rate above 35°C.

Growth analysis and daylength

Celosia is recorded as an obligate short-day plant. In trying to understand response to daylength in relation to season of growth and relevant temperatures, Fawusi *et al.* (1983) carried out controlled environment experiments to measure response to photosynthetically active radiation (PAR).

Results indicated that leaf fresh and dry weights at 28-day harvest increased with increasing PAR in both *Celosia* and *Amaranthus* (Table 4.2). Stem fresh and dry weights increased more with increasing PAR in *A. hybridus* than in *C. argentea.* Stem weights were similar in the two species at 215 μmol^{-2}s^{-1} but *A. hybridus* had about double the stem weight of *C. argentea* at 400 μmol m^{-2}s^{-1}. Leaf/stem ratios decreased markedly with increasing PAR in the former species but were similar at all PAR treatments in the latter.

Table 4.3 The effect of PAR on stem lengths (cm)

Species	*PAR* (*μmol photons $m^{-2}s^{-1}$*)	*Days from emergence* 7	14	21	28
Amaranthus	215	2.2 ± 0.2	2.9 ± 0.1	7.4 ± 0.5	17.7 ± 1.0
Green Head	325	2.0 ± 0.1	3.1 ± 0.1	8.7 ± 0.6	20.6 ± 1.1
	400	2.0 ± 0.1	3.5 ± 0.2	13.1 ± 0.5	22.8 ± 1.9
Celosia	215	2.6 ± 0.1	4.4 ± 0.3	7.0 ± 0.5	12.5 ± 0.5
IITA (SRV8)	325	2.6 ± 0.2	3.8 ± 0.1	6.2 ± 0.5	11.8 ± 0.8
	400	2.2 ± 0.1	3.4 ± 0.2	6.3 ± 0.4	12.4 ± 0.3

Means ± SE of two experiments each consisting of three plants per treatment

Stems elongated slowly from 7 to 14 days after emergence and more rapidly after 14 days (Table 4.3).

Initially, stem length was similar at all PAR treatments in both species, but it increased with increasing PAR in *A. hybridus* at 21 and 28 days. *C. argentea* plants had increased stem length at 21 and 28 days but there were no differences between the effects of PAR. At 28 days plants of the former species were substantially taller than the latter at all PAR treatments.

Leaf area increased with increasing PAR (Table 4.4). The two species differed in the onset and extent of leaf area response to the increasing PAR. At 14 days and 21 days the former species was much more responsive to increased PAR than was the latter. By 28 days plants of both species had a similar leaf area response to PAR.

A comparison of the growth components: relative growth rate (RGR), leaf area ratio (LAR) and net assimilation rate (NAR) for 7–14, 14–21, and 21–28 days from emergence intervals between harvests revealed differential effects of PAR in the two species with advancing plant age (Table 4.5).

Relative growth rate decreased with increasing age in both species, but

Table 4.4 The effect of PAR on leaf area (cm^2 $plant^{-1}$)

Species	*PAR* (*μmol photons $m^{-2}s^{-1}$*)	*Days from emergence* 7	14	21	28
Amaranthus	215	1.3 ± 0.2	30.7 ± 3.6	236 ± 24	949 ± 108
Green Head	325	2.6 ± 0.2	42.2 ± 3.9	369 ± 45	1263 ± 129
	400	3.5 ± 0.3	97.0 ± 13.1	595 ± 153	1954 ± 152
Celosia	215	3.2 ± 0.2	30.3 ± 1.2	218 ± 24	981 ± 98
IITA (SRVB)	325	2.7 ± 0.2	37.1 ± 2.1	289 ± 18	1274 ± 53
	400	3.8 ± 0.3	47.6 ± 6.4	418 ± 33	1804 ± 72

Means ± SE of two experiments each consisting of three plants per treatment.

Table 4.5 The effect of PAR on relative growth rate (RGR) (mg mg^{-1} $week^{-1}$), leaf area ratio (LAR) (cm^2 mg^{-1}) and net assimilation rate (NAR) (mg cm^{-2} $week^{-1}$), 14–21 and 21–28 days after emergence

		Days from emergence								
Species	(μmol photons m^{-2} s^{-1})	*7–14*			*14–21*			*21–28*		
		RGR	*LAR*	*NAR*	*RGR*	*LAR*	*NAR*	*RGR*	*LAR*	*NAR*
Amaranthus	215	3.90	0.643	6.07	2.30	0.392	5.87	2.03	0.250	8.12
Green Head	325	3.20	0.425	7.53	2.36	0.316	7.47	1.54	0.246	6.26
	400	3.84	0.391	9.82	2.48	0.217	11.43	1.11	0.162	6.85
Celosia	215	2.63	0.538	4.89	2.18	0.401	5.44	1.78	0.316	5.63
IITA (SRV8)	325	2.58	0.357	7.23	2.30	0.323	7.12	1.74	0.250	5.51
	400	2.74	0.347	7.90	2.38	0.281	8.46	1.57	0.240	6.54

more in *A. hybridus* than in *C. argentea*. It was somewhat higher in the former from 7 to 14 days but PAR had little effect on either species. At 14–21 days the RGRs of both species were similar but there was again no effect of PAR. In contrast, by 21–28 days the RGR of the former species decreased with each level of increasing PAR, whereas there was little PAR effect on the latter species. Leaf area ratio decreased with increasing plant age as did RGR but, in contrast to RGR, LAR generally decreased with increasing PAR in both species. Net assimilation rate increased with increasing PAR in both species during the 7–14 and 14–21 day periods with similar values for corresponding categories between the two weeks. However, the NAR for 21–28 days declined at 325 or 400 μmol $m^{-2}s^{-1}$ PAR for both species while increasing in *A. hybridus*, and not changing much in *C. argentea* at 215 μmol $m^{-2}s^{-1}$ PAR compared with 14–21 days.

Testing of Celosia *under long days*

Many useful tropical plants have never been tested in more temperate regions on the presumption that they will not survive or produce useful harvests.

Landraces cultivated by local farmers in some tropical African countries are very early flowering (14–20 days) with long narrow lanceolate leaves (NIHORT, 1987). The plants are predominantly green with few purple types which are rarely planted. A variety (TLV8), originally from Benin (R.O.), was introduced to Nigeria and found to be most promising in terms of yield and quality of leaves. The vegetative yield is 15–20 t/ha as compared to 5–8 t/ha for local landraces. TLV8 has large green leaves and its late flowering and maturing characteristics, as a leaf vegetable, make it quite suited for both harvesting by uprooting and cutting.

The author imported TLV8 from the NIHORT, Ibadan, Nigeria, to Ontario (Canada) to study performance and adaptability to summer long-day conditions. Seeds were sown in mid-June in 1991 and 1992. Leaf size was larger and flowering began as daylength decreased. Branching appeared more profuse and greater pigmentation developed. Leaf production was good and it appears that *Celosia* can be grown as a leafy vegetable in other climates. In Canada, early sowing would be needed for seed production.

Hence the adaptability of this vegetable has hardly been tested outside the tropics but it appears to have potential.

Daylength response of C. cristata

C. cristata was originally thought to be day neutral with respect to flowering. However, further studies have since determined that flowering in *C. cristata* is a quantitative short-day response. In one plumose and three *cristata* cultivars a photoperiod of 14 h or less resulted in early flower initiation compared with 16 h or continuous photoperiod. With an 8 h photoperiod, night interruptions with 3 h of incandescent light resulted in significant delay of flowering in two cultivars compared with plants that received an 8 h day extension. On the other hand, little or no difference in flowering time occurred between day extension and night interruption in two other cultivars. There were no differences in flowering time when the daylength was extended to 12 h by 4 h of incandescent or fluorescent lamps at equal illuminance. When 8 h light extensions were imposed to provide a 16 h photoperiod, there were no significant differences in flowering time between light sources for all but one cultivar (Piringer and Borthwick, 1961).

Driss-Ecole (1979) noted that when *C. cristata* seedlings were placed under photoperiods of 8 h or 12 h, flower initiation occurred by 40 days or when 7 to 12 leaves had initiated. Under a 16 h photoperiod 60 to 70 days were required for floral initiation to occur. Transferring plants from an 8 h to 16 h photoperiod after 30 days resulted in significantly higher percentages of flowering plants compared with transferring them after 20 days. However, all plants transferred from a 16 h to an 8 h photoperiod either after 20 or 30 LD reached anthesis by 70 days from sowing. These data support the hypothesis that activation of the meristem (meristem corpus) began under LD and was followed by activation of the axial tunica and formation of prefloral meristem under inductive SD. This supports the finding of an intermediate state of the meristem (état intermediére) under non-inductive photoperiods. Driss-Ecole (1979) also showed that some shift of cells in the meristem of *cristata* types from G2 to G1 phase of the mitotic cycle had occurred at the time of irreversible commitment to flower, similar to that in *Sinapsis alba*. By looking at the distribution of nuclear DNA, it was determined that the critical period for ontogenesis of the shoot apex was between 18 and 24 days.

Table 4.6 Root colonization by vesicular–arbuscular mycorrhizal (VAM) fungi

Host plant	*VAM characteristics*		*Rhizosphere soil characteristics*	
	Colonization (%)	*Mycorrhizal structures* [a]	*pH*	*Moisture (%)*
Northern Himalayan hills (soil temperature range 6–28°C, 15 cm deep)				
Celosia argentea	26.3	H, V	7.3	9.5
Celosia cristata	40.6	H	7.5	11.5
Indo-Gangetic plains (soil temperature range 6–36°C, 15 cm deep)				
Celosia cristata	4.5	H	7.3	13.0
Aravali Hills (soil temperature range 10–36°C, 15 cm deep)				
Celosia argentea	60.0	H, V, A	7.1	20.1

[a] H, hyphae; V, vesicles; A, arbuscules. Adapted from Shanker *et. al.* 1991.

The responsiveness of *Celosia* species to daylength might be modified by hybridization (see Section 4.6).

4.3.2 **Mycorrhizal symbiosis**

The species of the families *Amaranthaceae* are often considered to be non-mycorrhizal. This may not always be true, as there are infrequent reports of mycorrhizae occurring. Shanker *et al.* (1991) examined some species for a symbiotic association with mycorrhizae in soils of the Indian semi-arid and arid regions. Of 10 species representing five genera, all had mycorrhizae and vesicular-arbuscular mycorrhizal (VAM) spores in the rhizosphere soil, showing a uniform symbiotic potential from all geographical locations under highly variable edapho-climatic conditions (Table 4.6).

Most of the root segments had typically, thick aseptate, intermittently swollen, parallel hyphae forming 'H' and/or 'Y' connections. Vesicles were mostly terminal, globose, subglobose or elongated. Arbuscules were only seen in *Celosia argentea.*

VAM fungal species associated with *Celosia argentea* were *Glomus geosporum, G. aggregatus, G. pasciculatum* and with *C. cristata, G. aggregatum* and *G. lomus* sp. unidentified. VAM associations are important for use by the plant of soil phosphorus.

4.3.3 **Effects of drought**

Four indices of water status are often used by researchers to study its effect on physiology. They are leaf relative water content (RWC), sap osmotic potential, leaf tissue water potential and leaf vapour diffusion resistance (LDR). Adedeji (1984) studied the response of leafy *Amaranthus dubius* and *Celosia argentea* grown under conditions of field capacity and at limited water supply for 10 weeks after germination. In *Celosia*, at a given level of RWC, the values of LDR for young and old leaves were similar in the full water regimes but the LDR was greater for young than old leaves in the inadequate water regime. Under the latter, a decrease in leaf water potential occurred only if the RUC of the young leaves was appreciably lower than that of the old leaves.

Several researchers have claimed that larger decreases of leaf water potential per unit decrease of RWC, as observed in *Celosia* (and also *Amaranthus*) confer drought resistance. It is concluded that plants raised with insufficient water or limited water supply develop in ways that tend to restrict transpiration during subsequent dry or drought periods.

The occurrence of natural drought cycles and intervals between irrigations can have a more profound effect on the very leafy species *Celosia* or *Corchorus* than on many other cultivated species. With the objective of determining the effect on growth and development processes of withholding irrigation until a predetermined tissue water stress was reached, Fawusi *et al.* (1984) undertook experiments to study this aspect in Ibadan, Nigeria. In their study, this drying cycle was followed by ample irrigation to allow recovery from the stress; experiments were conducted in controlled environments to stabilize environment interactions, and conditions similar to those used for base-line growth studies. It appears that a cycle of water stress to a leaf water potential of -1 MPa had little deleterious effect in either *Celosia* or *Corchorus*. In *Celosia*, there was a triggering of axillary leaf development (without wilting and death of older leaves) that increased leaf number, fresh weight and area compared with well-irrigated control plants. The mild water stress suppressed apical dominance long enough to break dormancy of axillary buds. A water stress cycle -2 MPa had many more deleterious effects than did the -1 MPa cycle. All response variables were decreased to some extent in both species. Axillary leaf development was triggered in *Corchorus* plants after many leaves present at the beginning of the stress cycle had wilted and died. Axillary leaf development did not occur in *Celosia* plants exposed to this drying cycle, at least up to 7 days after the return to daily irrigation. *Corchorus* generally responded more rapidly to water stress than did *Celosia*. Growth retardation was evident by the fourth day in *Corchorus* but not until day 7 in *Celosia*.

The implications are that mild water stress has less effects on *Corchorus* than *Celosia* and in the former may enhance leaf production.

4.3.4 **Response to salinity**

There are no data relating to *C. argentea* but genotypes of *C. cristata* have been looked at. Sharma *et al.* (1989) studied the effect of salinity levels on physiological aspects of ten garden annuals, including *Celosia*, grown in both winter and summer seasons in soils of electrical conductivity (EC) 0,

3, 6, 9 or 12 mmhos/cm. They found that relative water content and leaf water potential generally declined as soil EC increased, but leaf diffusive resistance increased for both adaxial and abaxial leaf surfaces. They also investigated the effect of different salinity levels on mineral composition of the same ten garden annuals, including *Celosia*. Plants were assessed for their N, P and K contents. Plant N and P contents generally increased with increasing salinity levels whereas the K content decreased.

4.3.5 **Response to copper deficiency**

Copper deficiency may be widespread, occurring in many areas and with effects shown by many different plant species. Copper toxicity is less common, occurring either in limited areas of high natural availability of Cu or after the addition of Cu as a pollutant in sewage sludge or in mine wastes. In Nigeria, Cu toxicity is a common problem for the plants when they are cultivated in abandoned old cocoa farms where Cu-containing pesticides had been used over a long period of time.

Cu is one of the micronutrients necessary for growth and optimal development of plants. If Cu is the limiting nutrient in the presence of all other nutrients, the plant cannot complete its life cycle, yield and growth are progressively depressed, physiological processes are affected and the plant develops symptoms of deficiency. Also, if Cu is in excess, it inhibits the general metabolic activity, inducing deficiencies in other micronutrients or direct toxicity of the plant.

Adedeji and Farimokun (1984) reported changes in some growth parameters, micronutrients, chlorophyll and ascorbic acid concentrations in leafy *Celosia argentea* grown with concentrations of copper. Plant height, leaf area, dry weight per plant as well as ascorbic acid content and chlorophyll content were maximal at low concentrations. Higher concentrations were somewhat toxic with plant height reduced by up to 40%, but dry weight reduced more and ascorbic acid content less.

4.3.6 **Pigmentation**

Members of the *Amaranthaceae* are characterized by the presence of betacyanin pigments. Coloration of stems, petioles, or spots on the laminae and of inflorescences in *C. argentea* is predominantly affected by length of day and temperature at night; experiments by Adedeji and Farimokun (1984) showed phosphorus deficiency also had an effect.

In selection for green strains, it has to be remembered that any fall in night temperature can also trigger greater pigmentation, or relative lack of it needs to be part of the selection process.

The betacyanins are betanin, amaranthin and celosianin. Such secondary metabolites have not yet been studied in cell culture of *Celosia argentea*.

Table 4.7 Analysis of some leafy vegetables in Western Nigeria

Species	*Dry matter (%)*	*% of dry matter*			
		Crude protein	*Ether extract*	*Crude fibre*	*Ash*
Corchorus olitorius	13.1	22.79	2.96	6.4	9.40
Celosia argentea	10.8	27.47	4.20	7.20	47.68
Talinum triangulare	9.68	21.09	1.47	10.34	34.56
Amaranthus caudatus	9.1	29.70	2.69	9.23	19.33
Celosia laxa	6.9	21.03	2.48	6.92	40.20

4.4 BIOCHEMICAL COMPOSITION

Oke (1966) described the local use and cooking of leafy vegetables among the people living in the western part of Nigeria (i.e. the Yoruba-speaking States). After leaves have been boiled, they are made into soup. He estimated that individual leafy vegetable intake per meal amounted to approximately 50 g of fresh material, i.e. about 5 g dry material per meal. He added that it is not uncommon for vegetables to be taken twice a day and this habit is not restricted to any particular income group or district.

Oke gathered leaves from *Corchorus olitorius, Celosia argentea, Talinum triangulare, Amaranthus caudatus* and *Celosia laxa* over a three month period from plants of varying ages, so as to represent the types of sample that might constitute a meal, dried them at 100°C overnight and then determined contents of soluble and total oxalates, hydrocyanic acid, crude protein, crude fibre as well as ash, either extract and macro- and micro-elements, expressing his results as percentages of dry matter.

Reference to Table 4.7 shows that there were no major differences in the crude protein contents. All studied vegetables were rich in crude protein, and it could be inferred that a large proportion of the individual protein requirement could be met by these vegetables subject to a sufficiently high rate of intake. The method of preparing these vegetables varies from district to district; in some, after a preliminary boiling, the water is discarded; in others it is retained. Thus it is difficult to assess nutrient losses during preparation. Since nothing is known of the biological value of the true protein, valid conclusions about the importance of these vegetables in the diet cannot be made. On the average one can infer that at one meal approximately 1.25 g of crude protein would be consumed by an individual or double this quantity when vegetables are taken twice daily. Assuming that the true protein is about 1 g this would result in 1–2 g intake per day, and will represent from 2–5% of the daily protein requirement.

The ash contents are high (except that of *Corchorus*) compared with those of spinach and other vegetables. The important dietary constituents of the ash are calcium, phosphorus and iron. These vegetables can contribute a

Table 4.8 Calcium and phosphorus ratio and intakes per day of some Nigerian-grown vegetables

Species	*Ca/P*	*Intake in mg per day*	
		Ca	*Fe*
Corchorus olitorius	5.4	137	2.7
Celosia argentea	1.0	195	1.5
Talinum triangulare	7.0	141	1.9
Amaranthus caudatus	5.2	169	2.3
Celosia laxa	1.0	178	2.0

large proportion of calcium and phosphorus requirements, especially when these are used as a substitute for meat. The ash contents are also high compared with those of other common foodstuffs, e.g. yam 2.68%, cassava 5.23%, maize 3.68%, rice 1.39%, cowpeas 3.78%.

Table 4.8 shows the Ca:P ratios. There is a wide range from 1.0 to 7.0. Although the ratios are unbalanced (1:2 satisfactory), the vegetables are taken along with other dietary constituents so that satisfactory ratios could be met. The Ca:P ratios of the other major foodstuffs are yam 1:1.5, maize 1:5, cowpeas 1:3.3.

The Table also shows the dietary intake of calcium and iron assuming 50 g fresh weight are required per meal. The calcium figures will represent a range of 7–12% for an adult, 5–8% for pregnant or lactating women. The corresponding figures for iron will range from 8–20% for adults.

The Zn, Mo and Mn contents in Table 4.9 are high compared with other vegetables, and this may be attributed to the fact that the plants used for analysis were grown on soils derived from acid granites.

Sreeramulu (1982), Faboya (1983) and Tindall (1983) have reported more recent findings on *C. argentea* and *C. trigyna* mineral composition, and

Table 4.9 Minor elements in some leafy vegetables in Western Nigeria

Species	*% of dry matter*		*Fraction in p.p.m. of dry matter*						
	Ca	*Mg*	*Mn*	*Fe*	*Cu*	*B*	*Zn*	*Mo*	*Al*
Corchorus olitorius	1.72	0.35	94	340	19.5	31.9	60	7.4	343
Celosia argentea	2.44	1.71	182	193	27.1	23.6	87	10.8	382
Talinum triangulare	1.77	2.15	372	242	11.0	19.0	79	6.9	500
Amaranthus caudatus	2.12	1.55	314	287	19.5	44.0	108	9.6	390
Celosia laxa	2.23	2.49	297	249	10.2	20.2	75	9.0	508

Table 4.10 Representative values for nutrients per 100 g of edible portion

											Vitamins				
Species	*Parts of Plant* [a]	*Water (ml)*	*Calories*	*Protein (g)*	*Fat (g)*	*Carbohydrate (g)*	*Fibre (g)*	*Calcium (mg)*	*Phosphorus (mg)*	*Iron (mg)*	*β-carotene equiv. (μg)*	*Thiamine (mg)*	*Riboflavin (mg)*	*Niacin (mg)*	*Ascorbic acid (mg)*
Amaranthus spp.															
	leaves	85	48	5.0	0.7	5	1.5	250		4.0	1800	0.10	0.30	1.5	100
	leaves	84	42	4.6	0.2	8	1.8	410	103	8.9	5716	0.05	0.42	1.2	64
	leaves	87	43	4.4	0.4	5	1.3	230	55	5.0					
	leaves	89	26	3.6	0.1	4		154	74	2.9	6545	0.04	0.22	0.7	23
Celosia argentea															
	leaves	84	44	4.7	0.7	8	1.8	260	43	7.8					
Celosia trigyna															
	leaves	89	33	2.7	0.4	6	1.2	154	32	5.0	1925		0.10		10

[a] Four different analyses.

depending on the different methods of chemical analysis, plant age, environmental and edaphic factors and material provenance, levels of macro- and micro-elements appeared variable. In Tindall's compilation of the several analyses the water, calorie, protein and fat contents of *C. argentea* were comparable to those found in unspecified vegetable amaranths, but were higher than the latter with regard to carbohydrate, fibre, calcium and iron composition (Table 4.10).

C. trigyna had lower calorie, protein, fat, carbohydrate, fibre, calcium, phosphorus and iron amounts than *C. argentea* and the analysed species of amaranth. Lower β-carotene, riboflavin and ascorbic acid in *C. trigyna* were also evident.

4.4.1 **Mineral composition**

Further data on mineral composition can be derived from Faboya (1983) who collected a range of leafy vegetables from a farm in Ibadan, Nigeria. Data for *Celosia*, and comparisons with *Amaranthus* are shown in Table 4.11.

4.4.2 **Amino Acids**

Busson (1965) analysed material from Côte d'Ivoire and showed amino acid contents as in Table 4.12.

Table 4.11 Mineral contents of some green leafy vegetables. Modified from Faboya (1983)

Local name	*Species*	% *Moisture*	% *Total ash* (*dry basis*)	*Mineral content (mg g^{-1} dry weight)*		
				Na	*K*	*Ca*
Ewedu	*Corchorus olitorus*	90.6 ± 0.1	21.0 ± 1.0	0.11 ± 0.01	0.46 ± 0.01	0.41 ± 0.01
Ewedu	*Corchorus olitorus*	82.1 ± 0.2	16.0 ± 1.0	0.21 ± 0.01	0.36 ± 0.01	0.38 ± 0.01
Sokoyo-koto	*Celosia argentea*	93.3 ± 0.7	22.0 ± 1.0	0.33 ± 0.03	0.65 ± 0.01	0.42 ± 0.01
Sokoyo-koto	*Celosia argentea*	90.3 ± 0.6	22.0 ± 1.0	0.23 ± 0.02	0.69 ± 0.01	0.37 ± 0.01
Tete	*Amaranthus caudatus*	90.6 ± 0.3	24.0 ± 1.0	0.76 ± 0.01	1.28 ± 0.01	0.73 ± 0.01
Tete	*Amaranthus caudatus*	87.5 ± 0.3	20.0 ± 2.0	0.65 ± 0.01	0.86 ± 0.01	0.52 ± 0.01

Local name	*Species*	*Mineral content (mg g^{-1} dry weight)*				
		P	*Mg*	*Fe*	*Zn*	*Mn*
Ewedu	*Corchorus olitorus*	0.30 ± 0.01	1.76 ± 0.01	0.37 ± 0.02	0.041 ± 0.001	0.062 ± 0.002
Ewedu	*Corchorus olitorus*	0.28 ± 0.04	1.17 ± 0.04	0.53 ± 0.02	0.037 ± 0.001	0.082 ± 0.002
Sokoyo-koto	*Celosia argentea*	0.29 ± 0.01	0.93 ± 0.01	0.56 ± 0.01	0.046 ± 0.001	0.032 ± 0.002
Sokoyo-koto	*Celosia argentea*	0.18 ± 0.01	0.66 ± 0.01	0.41 ± 0.03	0.038 ± 0.001	0.036 ± 0.003
Tete	*Amaranthus caudatus*	0.31 ± 0.01	1.20 ± 0.02	0.43 ± 0.01	0.054 ± 0.001	0.069 ± 0.002

Table 4.12 Amino acid content of *Celosia argentea* and *C. laxa* leaves in Côte d'Ivoire (g/16 g N)

Amino acid	C. argentea	C. laxa
(dry weight basis)		
Arginine	4.2	4.4
Cystine	–	1.7
Histidine	1.6	2.1
Isoleucine	3.5	3.8
Leucine	6.4	6.6
Lysine	4.6	4.8
Methionine	1.5	1.6
Phenylalanine	4.2	4.7
Threonine	4.0	4.1
Tryptophan	–	–
Tyrosine	3.8	4.1
Valine	4.5	5.1
Aspartic acid	7.7	8.4
Glutamic acid	11.9	10.3
Alanine	4.4	4.5
Glycine	5.1	5.7
Proline	4.5	5.4
Serine	3.9	4.0

4.4.3 **Micronutrients**

Busson (1965) also analysed contents of macro- and micro-elements as shown in Table 4.13.

4.4.4 **Fatty acids**

Data are lacking on fatty acid composition of *Celosia* although Nam and Rho (1988) have reported linoleic, oleic, cerotic, palmitic and stearic acids in *Celosia cristata.*

4.4.5 **Antinutritional substances**

The presence of substances in tropical and temperate leaf vegetables which may be harmful to the consumer is well documented. The occurrence of cyanogenic glucosides in cassava, alkaloids such as solanine and solanidin in the Solanaceae, oxalate and nitrate in spinach (*Spinacia oleraces*) and/or saponin in many other vegetables and foods has been reported by many authors. The occurrence of these harmful substances is dependent on environmental conditions, crop fertilization programmes, cultivar type, age, post-harvest handling, various methods of preparation and cooking, etc.

Table 4.13 Macro- and micro-element composition of *C. argentea* and *C. laxa* leaves from Côte d'Ivoire

Element content on dry weight basis		
(a) *Macro-elements* (g/100 g)	C. argentea	C. Laxa
Ca	1.61	1.40
K	5.20	2.70
Mg	0.73	1.20
Na	0.03	0.04
P	0.27	0.27
(b) *Micro-element* (p.p.m.)		
Ag	<0.2	<0.2
Al	568	265
B	32	22
Ba	86	<0.3
Co	0.8	<0.1
Cr	4	1.2
Cu	9	2.1
Fe	484	330
Ga	<0.3	<0.3
Li	<0.3	<0.3
Mn	231	290
Mo	<0.2	<0.2
Ni	3	1.4
Pb	0.6	0.8
Rb	38	6
Sn	1.5	0.4
Sr	61	120
V	1.3	<0.2
Zn	66	72
(c) *Gross analysis* (g/100 g)		
Cellulose	11.3	16.4
Protein	28.8	23.4
Ash	19.9	12.5
Glucosides	35.6	43.9
Insoluble glucosides	35.7	38.1

Oxalic acid

Sadik (1971) gave an outline of the content of oxalate in some indigenous Nigerian leafy vegetables, including *C. argentea*. From Table 4.14 it is noted that a large percentage of the total oxalate in some vegetables was in a soluble form which could induce oxaluria if taken in excess. Tindall (1977) has remarked that it would therefore be necessary to pursue further studies

Table 4.14 The oxalate content of some leafy vegetables from the Ibadan–Ilesha area analysed at IITA (from Sadik, 1971)

	Oxalate (% dry wt.)		
	Total	*Soluble*	*Insoluble*
Amaranthus hybridus	14.33	5.60	8.73
Talinum triangulare	14.21	13.37	0.84
Basella alba	12.94	4.52	8.42
Celosia argentea	11.92	7.55	4.37
Basella rubra	11.14	6.59	4.55
Manihot esculenta	5.29	2.66	2.63

into the residues of oxalates left in leafy vegetables following various methods of preparation and cooking, not to mention the effects of culture practices such as nitrogen fertilization (rate, form, application date, etc.), environmental factors and plant age on the oxalate content.

Adeyemi (personal communication, 1991), Director of the National Horticultural Research Institute, Ibadan, remarks that the presence of oxalic acid in vegetables of the families Amaranthaceae, Basellaceae and Portulacaceae is an inherent characteristic. He cites that oxalic acid becomes detrimental when the amount exceeds the amount of calcium in the vegetable because the excess oxalic acid binds other available calcium in the diet, forming an insoluble calcium oxalate in the digestive tract and possibly stones in the blood tract (oxalaemia) or in the urinary tract (oxaluria). However, blanching or boiling removes about 25–50% of the oxalic acid content (Oke, 1968; Taylor, 1979), and the lethal dose is actually reduced by the normal mode of vegetable preparation in Nigeria.

Nitrate in leafy greens

While nitrate *per se* is not harmful to man, nitrate–nitrites are considered precursors of carcinogenic nitrosoamines. The effect of nitrite ion is to

Table 4.15 Concentration of nitrate in some vegetables grown in the Benin area

Botanical name	*Number of samples*	*Moisture content %*	*Nitrate p.p.m. dry wt. (mean ± s.d.)*
Corchorus olitorius	7	72	560 ± 69
Basella alba	8	94	764 ± 95
Brassica oleracea	7	86	1114 ± 55
Celosia argentea	11	85	1281 ± 85
Talinum triangulare	8	92	1345 ± 242
Amaranthus hybridus	8	87	1675 ± 116

oxidize haemoglobin to methaemoglobin. It is also known that the nitrate–nitrite content will vary and depend on the amount of N-fertilizer applied, nitrogen form, plant age, environmental and edaphic factors.

Sanni (1983) assessed the nitrate–nitrite in some common leafy vegetables found in Benin city environs (Nigeria). Sanni carried out his analysis on representative samples collected from local markets, measuring the nitrate content using an Orion research digital Ion Analyser model 701A and estimating the nitrite content by difference between the nitrate content before and after oxidation (Table 4.15).

No nitrate was detected in any of the vegetables analysed so in order to investigate the nitrite recovery, samples of four specimens were labelled with 10 p.p.m. and 20 p.p.m. nitrite solution, and the nitrite content was thus determined. There was a good recovery (96–101%) of nitrite in the vegetable samples with added nitrite.

With the increase in the number of working housewives in Nigeria, there is an increase in storage of vegetables in refrigerators prior to use in meals. The nitrate in these vegetables may be converted into nitrite due to microbial action during storage, leading to an increase in the nitrite level. Even after storage in a refrigerator for 48h, there was no appreciable loss of nitrate.

4.4.6 **Secondary compounds**

Data on ornamental *Celosia* species have indicated the presence of Δ^7-sterols. No experimentation has been conducted using edible vegetable *Celosia* species.

Tests using *Celosia argentea* for insecticidal and repellent constituents have been inconclusive. However, traditional use of *Celosia* as a vermifuge and other medicinal purposes may be borne out by more detailed investigation.

4.5 AGRONOMY

Although widely grown, most production is in backyard systems, but enough is known to make clear recommendations on agronomy needs.

First of all the crop requires a fairly high level of soil fertility and normal seed bed preparation. *Celosia* (and amaranth) respond to adequate fertilizers with a 3–4 fold increase in yield. On poor and moderately fertile soils up to 400 kg/ha NPK(10.10.20) is recommended, often mixed with organic manure or even town refuse. Split application of fertilizer, especially in the rainy season, is useful on poor soils, half during seed bed preparation, the other half during the growth period. It can be applied either in granular form or dissolved in irrigation water. Grübben (1975) showed how successive crops with fertilization converted poor sandy soils into good garden soils.

Spacing varies between 20–100 plants per m^2, depending on harvesting. Plots can be devised if harvesting is by uprooting in a relatively short period, as short as 30 days. If repeated cutting is practised density is lower at *ca.* 30 plants per m^2.

In repeated cutting it is essential that attention is paid to timing and height of the cut. Twenty-day intervals are normal, with cuts at *ca.* 20 cm above ground. Cutting higher up the plant can precipitate early flowering and cutting lower may cause poor regrowth.

In West Africa, wet-rot fungus *Choanephora cucurbitarum* (Berk and Rav.) Thaxter can cause damage especially in dense plots and when transplanting is practised and also at times after cutting. Additionally damping-off by *Pythium aphanidermatum* (Eds.) Fitzp. may occur. The former is controlled by plant density, the latter by good cultivation and sowing at *ca.* 3 g seed per m^2 of seedbed with thinning.

Seeds lose viability rapidly under hot, humid conditions. They need to be cleaned and kept as dry as possible in cotton bags, particularly in air-conditioned rooms if possible.

4.5.1 **Seeding and general cultural practices**

Normal sowing at about 3g per m^2 potentially produces 3000 seedlings. Reduction in waste through thinning should be a local aim of growers.

Tindall (1983) described the traditional cultural techniques of cultivating the green and red forms of *C. argentea* and their yields in tropical West Africa. For leaf production of *Celosia* green form, seeds are sown broadcast in seedbeds which may be temporarily mulched with dried grass which is removed after germination. The seedlings are harvested by thinning out the tallest plants after they reach a height of 15 cm which may be 30–40 days from sowing. Harvesting continues until the plants are about 30 cm in height. In leaf production of the red form, seeds are sown broadcast and transplanted when 10–15 cm high to prepared beds at a square planting of 15–30 cm apart each way. The plants are harvested when they are about 30 cm tall, initially by thinning out closely planted seedlings up to 25–30 cm high, but later by removing the terminal shoots from the remaining plants. This encourages lateral branching and provides shoots and leaves for successional harvests. Approximate yields reported in Nigeria were: green form, 8 kg from 5 m^2 (16 t/ha); red form, 14 kg from 5 m^2 (28 t/ha).

In large-scale production, the planting method, plant population and harvesting practices will indeed vary from the above traditional methods. More organized investigations conducted by the National Horticultural Research Institute (NIHORT), Ibadan (Nigeria) showed that transplanted *Celosia* seedlings gave 3–4 times better yields than when seeded. A good crop of *Celosia* is obtained when 2 to 3-week-old seedlings are transplanted, and that in case of direct seeding, 6–8 g/ha of seeds is required for optimum

Table 4.16 Yield and yield components of *Celosia* harvested by cutting and uprooting in tropical Nigeria (NIHORT, 1987)

	Harvests					
	1	*2*	*3*	*4*	*5*	*Total*
Cutting (15 × 15 cm)						
Shoot fresh wt. (t/ha)	16.76	17.70	14.25	7.06	3.00	58.77
Leaf fresh wt. (t/ha)	12.96	11.50	10.14	5.99	1.76	42.35
Leaf area (cm^2)	203 128	59 129	46 313	15 470	6630	330 670
Uprooting (at 10 × 10 cm)						
Shoot fresh wt. (t/ha)	24.43	15.97	8.21	-	-	48.61
Leaf fresh wt. (t/ha)	10.69	5.55	3.82	-	-	26.06
Leaf area (cm^2)	146 683	102 051	73 341	-	-	322 075

seedling establishment and vegetative yield of the crop. Comparative trials of the two methods of harvesting (by uprooting or repeated cutting) showed that transplanted crop of *Celosia* performed best when harvested by cutting at a spacing of 15 × 15 cm. In the case of harvesting by uprooting, a spacing of 10 × 10 cm was found adequate for optimum yield (Table 4.16).

The yield of *Celosia* in these shortday tropical conditions appeared to be affected by the cutting interval. The best time for harvesting is at three-weekly intervals and at a height of 20 cm above the ground (Table 4.17). In a more recent study carried out at NIHORT (Fasheun, 1989), four cultivation methods of *Amaranthus hybridus* and *C. argentea* were compared: no-tillage, flat beds, ridges and raised beds. It has been found that soil cultivation increased growth rates and yields by 74% in *A. hybridus* and 50% in *C. argentea*. It was concluded that flat beds were better than ridges or raised beds since they conserved water and had a high infiltration rate.

In York (Ontario, Canada), *C. argentea* line TLV8 gave promising vegetative performance when sown 20 cm apart on flat beds in the summer seasons of 1991 and 1992 (Badra, unpublished). Influenced by the climatic data recorded in summer 1991, *C. argentea* plants were vegetatively harvestable between 39 and 108 days after sowing. Similarly, stands pinched above the second node regenerated and produced lateral shoots with fresh leaf

Table 4.17 Effect of cutting height and frequency of cut on vegetative growth of *Celosia argentea* (kg/m^2) in tropical Nigeria (NIHORT, 1987)

	Frequency of cut		*Height (cm) of cut*		
	2 weeks	*3 weeks*	*10*	*15*	*20*
Marketable yield	37.71	**69.68**	49.49a	53.12a	**59.95**
Leaf yield	23.13	**28.86**	24.89a	24.99a	**28.13**

produce for the consumer. Consumer's acceptability under these climatic conditions lasted through the latter plant age.

Mixed cropping

As outlined earlier, it is not an uncommon practice in the tropics for vegetables to be grown in mixtures, especially for domestic purposes. Large-scale monocropping seems to be the case in commercial and/or government- or institution-run operations. Leafy vegetables are no exception to the traditional practice whereby different food crops such as maize, yams, cassava, capsicum peppers, cowpeas, groundnut, okra, tomato, cocoa, etc. are interplanted to meet domestic consumption. In these situations, a produce surplus to the family need is either traded or marketed locally.

Undoubtedly, the above situation gives room to competition factors which could influence the involved species positively or negatively. Various authors have shown that the yields of vegetables grown in mixtures with root crops were reduced by as much as 50% of their sole crop equivalent in Nigeria. In this country, leafy *Amaranthus*, *Celosia* and *Corchorus* are often intercropped because they are indispensable in most soup and salad preparations. Suspecting that the vegetable yields from such species mixtures would be higher than in cereal/vegetable or root crop/vegetable mixture, Ikeorgu (1990) investigated the productivity of a mixture of the above-mentioned leafy species when compared with their sole crops.

Ikeorgu's results indicated that amaranth gave the highest vegetable and dry matter yield when intercropped with both *Celosia* and *Corchorus*. Amaranth alone gave the lowest vegetable and dry matter yield.

Vegetable and dry matter yield of *Celosia* were highest under sole cropping and lowest in *Celosia*/*Amaranthus* intercrop, suggesting that amaranth could be suppressing *Celosia* in a 2-crop mixture.

Bhaskar *et al.* (1984) studied *C. argentea* mixtures with food legume crops and when dense enough for the *Celosia* to be a weed found interference with nitrogen relations of the legumes with inhibition of nitrogen-fixing bacteria. Interplanting with cowpea, Lima bean, groundnut or bambara groundnut - common in West Africa - or with kidney bean, green gram, black gram or bambara groundnut - common in Tanzania - may mean that the *Celosia* benefits from the association.

Normally weeds are slashed and allowed to decompose *in situ* and serve as a mulch. There has been increasing evidence both from germination tests and from the field that plant residues left in the soil liberate a number of organic compounds which are inhibiting to succeeding crops. In Ibadan (Nigeria), like other areas in the humid tropics, luxuriant weed growth is encountered immediately after the first rain in March/April, and by the time of land preparation weed communities characteristic of the area are fully established. Saleem and Fawusi (1983) undertook greenhouse and laboratory studies to evaluate the effects of tropical weed residues on seed germination and seedling growth of tomato, capsicum pepper, *C. argentea* and sorghum. Involved in their trials were three environmentally-predominant weed species, namely *Euphorbia heterophylla*, *Cyperus rotundus* and *Digitaria sanguinalis*. They found that tomato germination was markedly reduced but the effects on *Celosia* germination were variable as was the longer-term effect on seedling dry matter production. In areas where mulching from weeds is widespread, tillage is recommended to avoid potential problems of residues.

Most weeds can be well controlled through weekly weeding, except for *Cyperus*. In this case it is necessary to destroy tubers using suitable herbicides.

4.5.2 **Pests**

Apart from those affecting emergence and plots after plant cutting mentioned above, two important pests need attention: (1) caterpillars, *Hymenia reculvalis* F.; and (2) nematode, *Meloidogyne incognita* Chitwood, the latter being particularly serious on *Celosia*.

Control of the former is simple with a range of insecticides, control of the latter more difficult due to build up in the soil. Control requires a mix of soil fumigation and rotation of *Celosia* with other vegetables not affected. However, nematode populations can show speedy recovery (Grubben, 1975).

A range of other locally important pests are:

Alcidodes albolineatus Boh., S.W. Nigeria;
Anacatantops notatus Kars, S.W. Nigeria;
Aspavia armigera F., West Africa;
Cassida enrviz Boh., India;
Cletus fuscescens Wlk., West Africa;
C. unifasciata Blote, West Africa;
Gasterochisus rhomboidalis Boh., West Africa;
Lagria villosa F., Nigeria;
Myzus persicae Sulz., Nigeria;
Nysius albidus Dall, Nigeria;
Polyphagotarsonemus latus (Banks), Nigeria;
Sjostedtina sp., Nigeria;
Zonocerus variegatus L.

4.5.3 **Fungal diseases**

Some fungi have been reported on *C. argentea* causing slight, moderate or severe damage in the humid tropics. When infected, leaves look undesirable for consumption and with severe attacks, leaves reach senescence faster than uninfected leaves and reduced yield of the foliage occurs at harvest. Diseases such as rusts and mildews occur predominantly in the dry seasons while the blights and rots occur more on plants in wet seasons. The leaf spotting diseases are found to occur in severe proportions when plants are growing with minimum water supply. The fungi are:

Albugo blitii, powdery mildew, serious in West Africa;
Alternaria sp., leaf spot, moderate damage;
Cercospora celosiae, leaf spot, moderate to minor damage;
Choanephora sp., leaf blight, prevalent in hot humid weather and after insect damage;
Colletotrichum sp., leaf spot, minor importance;
Curvularia sp., leaf spot, slight damage in S.W. Nigeria;
Fusarium lateritium, fusarium wilt, moderate damage in Nigeria;
Thatatephorus cucumeris, damping off, widespread.

4.5.4 **Viruses**

Smith (1972) included *C. argentea* in his host range and susceptibility studies of some virus diseases. These were:

1. beet western yellow virus;
2. beet curly top virus;
3. carnation Italian ringspot virus;
4. carnation mottle virus;
5. carnation vein mottle virus;
6. poplar mosaic virus;
7. potato leafroll virus and
8. raspberry bush dwarf virus.

In prevention and control, sickly plants should be totally uprooted and burned, farm sanitation maintained through regular weeding, and if necessary chemicals used to control populations of possible insect vectors (grasshoppers, aphids, thrips, etc.).

4.5.5 **Yield**

From experimentation with local cultivars with promising qualities, Omueti (1980) partitioned yield into marketable yield and edible yield. The former is defined as the tops of the plant as they are harvested from the field and sold in the market, the latter is defined as the leaves a housewife would pick to use as a pot vegetable. Yield is shown in Table 4.18.

Both marketable and edible yields increase with age, the rate of increase of the former being higher than the latter. Farmers can realize higher economic returns by harvesting at later stages of growth but nutritional

Table 4.18 Yield component of *Celosia* (mean of six ages)

Cultivar	*Marketable yield (kg/m²)*	*Edible yield (kg/m²)*	*Leaf/stem ratio*	*Dry matter content (%)*
V1	7.3	2.9	0.0	8.7
V2	8.3	0.8	0.6	12.5
V3	6.4	2.4	0.9	9.4
V4	6.1	2.2	0.9	9.5
SE	0.3	0.1	0.5	0.2

value may not be the best. There are genetic differences in protein and vitamin C contents, depending upon the cultivar.

4.6 GENETICS AND BREEDING

4.6.1 **Germplasm collections**

Although materials were collected during the project referred to in Section 4.1, there are no comprehensive, fully documented and duplicated collections. According to IBPGR records there are also a few samples in Japan, (National Institute of Agrobiological Resources, Tsukuba), South Africa (Directorate of Plant and Liquor Control, Pretoria) and USA (National Plant Germplasm System). These limited samples are 'curiosities'. There is an urgent need to establish collections of primary diversity from Africa and secondary diversity from the Indian subcontinent (although some collecting has been done in India) and Southeast Asia and for the IBPGR to determine the status of the collected materials in Nigeria (Badra, 1985). There is a possibility that other samples exist in other collections mixed with amaranths.

Descriptors and descriptor states for documentation of germplasm of *Celosia* have been developed. These cover passport information (standard collection site description), characterization (a limited number of morphological characteristics) and further evaluation (descriptors related to use).

4.6.2 **Cytology**

The base number of chromosomes is $x = 9$. Chromosome numbers of *C. argentea* are $2n = 36, 72$ and of *C. trigyna* $2n = 18$. The former is tetraploid and octoploid, the latter diploid.

C. cristata, the garden cockscomb with a fasciated inflorescence, also has chromosome number of $2n = 36$ and the cultigen var. *plumosa* which arose in cultivation is also $2n = 36$. In all possibility *C. cristata* arose from the wild tetraploid cytotype of *C. argentea* (Khoshoo and Pal, 1972).

The tetraploid cytotype of *C. argentea* is morphologically very heterogenous and wide diversity is apparent in leaf shape and size and also in branching pattern. The tetraploid and octoploid cytotypes when hybridized produce almost completely sterile F1s but a few seeds can be formed which result in aneuploids ($2n = 39–41$) and 12-ploids ($2n = 108$). The latter are fertile and are relatively easily established. The distribution of cytotypes in nature has not been fully investigated (Pal and Khoshoo, 1983).

The ornamental forms of *C. cristata* are fully fertile with *C. argentea*; the chromosomes in the hybrids pair well at metaphase I followed by a normal meiotic course resulting in high pollen and seed fertility. Interspecific hybridization could be important in widening adaptability.

The breeding of *C. argentea* can be broadened by the use of wide crossing. *C. whiteii* Grant, $2n = 108$, from Malaysia and S. India is a natural allopolyploid, probably derived from *C. argentea* and *C. cristata*, but is slower growing than *C. argentea* even though the species has individuals with fasciated or non-fasciated inflorescences. Many species are virtually 'unknown' (Appendix).

Further information on chromosome content of wild *Celosia* species of Africa and S. India–Sri Lanka would be helpful in designating primary and secondary genepools of the cultigen. Also *C. nitida* L. of Guyana, Martinique and neighbouring areas would repay study. Almost certainly, resistances to major diseases and pests, e.g. wet rot or nematodes, would be found in wild species. Additionally, the ornamental species produce larger foliage and this trait could well be useful in *C. argentea* (Pal and Khoshoo, 1983).

4.6.3 Breeding and selection

Although wide crossing is potentially useful, breeding is likely to continue through selection from local cultivars. These are known to show major differences in leaf/stem ratios, leaf yields, crude fibre content, crude protein of leaves and vitamin C contents.

For instance a cultivar collected in Benin by the International Institute for Tropical Agriculture (IITA) when introduced to Ibadan, Nigeria, yielded a 20–25 t/ha compared to 5–8 t/ha from local farms.

Selection will be more rapid if a comprehensive germplasm collection, well documented, is fully available.

4.7 POTENTIAL

Recent decades have shown an increased awareness that leafy vegetables can provide a good proportion of dietary proteins, minerals and vitamins especially in tropical areas. The last decade has shown a number of experimental studies on *Celosia* so that there is a much better understanding of its growth, nutrition, and place in farming systems.

In the tropics, introduction of exotic vegetables tends to push aside the use of local vegetables, but there is increasing understanding that the latter have a continuing role to play especially in managing rural ecosystems in a sustainable way.

Amaranths, as a result of research since the mid-1970s, have become widely known. In part this has been because amaranths can be used for seeds as well as leafy vegetables, but the key to amaranth success is probably the widespread sharing of germplasm between continents and resultant contributing research in more than one continent. This has hardly happened with *Celosia*, although as will be seen from the account above, *Celosia* and *Amaranthus* (and sometimes *Corchorus*) complement each other in tropical backyards. In part also *Celosia* is equated with the hot, humid tropics and little research has been done elsewhere, in contrast to the work on amaranth.

However, sufficient research has been done to develop and implement better technologies for the culture of *Celosia*. These should be adapted to farming systems research in suitable areas so that the potential of higher yields and more nutritious forms can be realized. Technology is at present 'on the shelf' but should be applied.

APPENDIX

Some relatively unknown *Celosia* species

Some examples are listed below:

Celosia alomiris Sessé and Moc, Mexico
Celosia anthelminthica Aschers, Africa
Celosia baronii Cavaco, Madagascar
Celosia brevispicata
Celosia chiapensis Brandeg, Mexico
Celosia digynia Sessé and Moc, Zimbabwe and Tanzania
Celosia dioica Sessé and Moc, Mexico.
Celosia exellii Suesseng, Angola.
Celosia humbertiana Cavaco, Madagascar.
Celosia monosperma Rosé, Mexico
Celosia orcuttii Greenma, Mexico
Celosia paniculata L.
Celosia persicaria Schinz, Peru
Celosia populifolia Moq, Mainland Africa
Celosia salicifolia Lopr, Brazil
Celosia spicata (Thouars) K. Spreng, Madagascar
Celosia triquetra Sessé and Moc, Mexico
Celosia virgrata var. *paniculata* Suesseng, Peru and Ecuador.

It is highly likely that the genus *Celosia*, rather in parallel with *Amaranthus*, shows several areas of high diversity. A scientific study is urgently needed and also elucidation of genetic affinities between *Celosia* and *Lugrezia*, *Hermbstaedtia* and other genera.

REFERENCES

Adedeji, F.O. (1984) Comparison of studies of plant water status in two (*Celosia argentea* L. and *Amaranthus dubius* Mart. ex Thell.) tropical leaf vegetables. *Turrialba*, **34** (3), 387–91.

Adedeji, F.O. and Farimokun, V.O., (1984) Copper deficiency and toxicity in two tropical leaf vegetables (*Celosia argentea* L. and *Amaranthus dubius* Mart. ex Thell.) *Environment and Experimental Bot.* **24**, 105–10.

Allison, E.C. (1989) *The Besler Florilegium Plants of the Four Seasons*. H.N. Abrams Inc., New York, N.Y.

Badra, T. (1985) Passport data of horticultural germplasm collections made in 1983 and 1984. Tech. Comm. NIR/84/014, FAO, Rome, Italy.

Berhaut, J. (1971) *Flore Illustré du Senegal*, I, Clairafique, Dakar, Senegal.

Bhaskar, A., Vyas, K.G. and Pandya, S.M., (1984) Nitrogen relationships of *Vigna mungo* in five stands and in *Celosia argentea* mixtures. *Ann. Bot.*, **54**, 799–805.

Burkill, H.M. (1985) *The Useful Plants of West Tropical Africa*, ed. 2, Vol 1, Families A–D. Royal Botanic Gardens, Kew, UK.

Busson, F. (1965) *Plantes Alimentaires de l'Ouest Africa*, Le Comte, Marseille, France.

Dalziel, J.M. (1937) *The Useful Plants of West Tropical Africa*, Crown Agents, London.

Driss-Ecole, D. (1979) Etude de la tesseur en DNA nucleaire dans la meristem apical des tiges du *Celosia cristata* (Amarantacées) en fonction de la photopériod. *Can. J. Bot.*, **57**, 2760–5.

Eliasson, U.H. (1988) Floral morphology and taxonomic relations among the genera of Amaranthaceae in the New World and the Hawaiian Islands *Bot. J. Linnean Soc.*, **96**, 235–83.

Faboya, O.O.P. (1983) The mineral content of some green leafy vegetables commonly found in the western part of Nigeria. *Food Chemistry*, **12**, 213–16.

Fasheun, A. (1989) The effects of soil and micro-climatic environment on the field performance of *Amaranthus hybridus* L. and *Celosia argentea* L. *Res. Bull. Nat. Hort. Res. Inst.*, Ibadan, **6**, 1–5.

Fawusi, M.O.A., Ormrod, D.P. and Eastham, A.M. (1983) Effects of different levels of photosynthetically active radiation on the growth of *Amaranthus hybridus* and *Celosia argentea*. *J. Hort. Sci.*, **58**, 555–9.

Fawusi, M.O.A., Ormrod, D.P. and Eastman, A.M. (1984) Response to water stress of *Celosia argentea* and *Corchorus olitorius* in controlled environments. *Scientia Horticulturae*, **22**, 163–71.

Grubben, G.J.H. (1975) La culture de l'amarante légume-feuilles tropical. *Meded. Landbouwhogeschool, Wageningen*, **76** (6), 1–222.

Grubben, G.J.H. (1977) *Tropical Vegetables and their Genetic Resources*. International Board for Plant Genetic Resources, Rome, Italy.

Harvard University (1968) *Gray Herbarium Index*, Vol. 2, G.K. Hall, Boston, MA.

Hutchinson, J., Dalziel, J. and Keay, R.W.J. (1966) *Flora of West Tropical Africa* 2nd ed., Vol I, part 1, Whitefriars Press, London.

Ikeorgu, J.E.G. (1990) Glasshouse performance of three leafy vegetables grown in mixtures in Nigeria. *Scientia Horticulturae*, **43** (3–4), 181–8.

Khoshoo, T.N. and Pal, M. (1972) The probable origin and relationships of garden cockscomb. *Bot. J. Linnean Soc.* **66**, 127–41.

Kumar, V. and Singh, J.N. (1985) Efficiency of herbicides for weed control in curdbean (*Vigna mungo*). *Abstr. Papers Ann. Conf. Indian Soc. Weed Sci.*, 55–6.

Marnotte, P. (1984) Influence of agro-ecological factors on weed development in a humid tropical climate. *Comptes Rendu du 7iéme Colloque International sur l'Ecologie, la Biolgie et la Systematique des Mauvaises Herbes*, **1**, 183–90.

Nam, H.K. and G.H. Rho (1988) Composition of fatty acids and amino acids in the aqueous extract of cocks' comb roots. *J. Korean Soc. Food and Nutrition*, **17**, 172–5.

NIHORT (1987) *Annual Report 1986*, National Horticultural Research Institute, Ibadan.

Oke, O.L. (1966) Chemical studies on some Nigerian vegetables, *Tropical Science*, **8**, 128–31.

Oke, O.L. (1968) Chemical changes in some Nigerian vegetables during growth. *Expl. Agric.*, **4**, 345–6.

Okusanya, O.T. and Ahmedu, M. (1983) Competition between *Celosia cristata* and *Corchorus olitorius* under different environmental conditions. *Tropical Ecology*, **24**, 113–21.

Omueti, O. (1980) Effect of age on some *Celosia* cultivars. *Expl. Agric.*, **16**, 279–86.

Pal, M. and Khoshoo, T.N. (1983) Utilization of resources of *Celosia argentea. Plant Genetic Resources Newsl.*, **53**, 22–4.

Piringer, A.A. and Borthwick, H.A. (1961) Effects of photoperiod and kind of supplemental light on growth, flowering, and stem fasciation of *Celosia. Am. J. Bot.*, **48**, 588–92.

Rajan, M.S.S., Rao, R.S. and Sesliaiah, T. (1984) Weed flora in groundnut as influenced by herbicides. *Madras Agric. J.*, **71**, 697–8.

Rao, A.N. Kondap, S.M. and Murty, Y.V.N.(1982) Weed spectrum and pod yield of groundnut as influenced by phosphorus levels and weed management. *Abstr. Papers Ann. Conf. Indian Soc. Weed Sci.*, 26.

Sadik, S. (1971) Oxalate contents of some leafy vegetables. *Ford Foundation/IITA/IRAT Seminar on Vegetable Crop Research*, Ibadan, Nigeria.

Saleem, M.A.M. and Fawusi, M.O.A. (1983) A note on the effects of tropical weed decomposition on seed germination and seedling growth of some agricultural crops. *Agriculture, Ecosystems and Environment*, **10**, 347–52.

Sanni, S.B. (1983) A survey of nitrate–nitrite in vegetables in Benin environs. *J. Plant Foods*, **5**, 75–9.

Shanker, N.A., Mathew, J. and Varma, A. (1991) Occurrence of vesicular-arbuscular mycorrhizae with *Amaranthaceae* in soils of the Indian semi-arid region. *Biology and Fertility of Soils*, **11**, 140–4.

Sharma, S.S., Yamalagni, R. and Arora, R.K. (1989) Effect of different salinity levels on mineral composition of garden annuals. *South Indian Hort.* **37**, 291–4.

Shelke, D.K., Bhosle, R.H. and Jadhar, N.S., (1985) Studies on integrated weed management in sorghum-pigeonpea microcropping system. *J. Maharashtra Agric. Univ.*, **10**, 233–4.

Singh, J.P., Rose, H.S. and Sangita (1985) Biology of *Cassida enervis* Boh. (Coleoptera: Chrysomelidae: Cassidinae) a serious pest on *Celosia argentea* an ornamental plant. *Entomol.*, **10**, 107.

Smith, K.H. (1972) *A Textbook of Plant Virus Diseases*, 3rd edn, Academic Press, New York.

Sreeramulu, N. (1982) Chemical composition of some green leafy vegetables grown in Tanzania. *J. Plant Foods*, **4**, 139–41.

Taylor, G. (1959) *Index Kewensis*, Suppl. 12, Royal Botanic Gardens, Kew.
Taylor, O.O.A. (1979) *Influence of Age and Nitrogen Fertilization on the Nutritive Value of Five Nigerian Leafy Vegetables*, Ph.D. thesis, University of Ibadan, Nigeria.
Tindall, H.D. (1977) Vegetable crops, in *Food Crops of the Lowland Tropics*, (eds C.L.A. Leakey and J.B. Wills), Oxford University Press, pp. 101–25.
Tindall, H.D. (1983) *Vegetables in the Tropics*, Macmillan, London.
Townsend, C.C. (1973) Notes on Amaranthaceae I. *Kew Bull.*, **28**, 141–6.
Townsend, C.C. (1982) A new species of African *Celosia* and a new conspectus of *Hermbstaedtia*. Notes on Amaranthaceae XIV. *Kew Bull.*, **37**, 81–90.
Usoroh, N.J. (1989) Weed control in horticultural crops. *Occasional Papers, NIHORT*, Ibadan, No. 18, 76–85.

Part Two

Promising Crops Deserving More Research Attention

5

Vegetable jute (*Corchorus*)

R. I. Denton

5.1 HISTORY, TAXONOMY AND ECOLOGY

The family *Tiliaceae* comprises about 40 genera and 400 species of trees, shrubs and herbs largely distributed in warm and tropical regions. The genus *Corchorus* includes 50–60 species which are extremely variable, but all species are apparently highly fibrous of which two species, *Corchorus capsularis* L. (jute, white jute) and *C. olitorius* L. (nalta jute, tossa jute, tussa jute, Jew's mallow) are major world economic bast fibre crops of particular importance to the economies of Bangladesh, India, China and Nepal where the bulk of the world's supplies are grown (Edmonds, 1990; Purseglove, 1968). A secondary but extensive use of some of the herbaceous annual *Corchorus* species, particularly *C. olitorius*, is their cultivation as a local leaf spinach-type vegetable both in the jute (fibre) producing countries and also in hot moist lowland tropical and sub-tropical areas worldwide. Such types are shorter and more branched than the fibre crop forms which may reach 5 m in height (Westphal-Stevels, 1986).

Kundu (1951, 1956), and Kundu *et al.* (1959) have discussed the origin of jute, stating that the primary centre of origin of *C. olitorius* is Africa, and the secondary centre may be India or Indo-Burma, although it is now pantropical in distribution. This species has since become naturalized in most tropical regions, as an escape from cultivation. The primary centre of origin of *C. capsularis* is the Indo–Burmese region, which according to Vavilov's concept includes south China. It is not found in Africa or Australia. Earlier accounts suggested, on the basis of the occurrence of wild species, that the centre of origin was China but the wider area now prevails.

Jute has been primitively cultivated and used since ancient times in Africa and Asia to provide a cordage and weaving fibre from the stem and food from the leaves and tender shoots. Full domestication and development of

Pulses and Vegetables. Edited by J. T. Williams. Published by Chapman & Hall,
2–6 Boundary Row, London SE1 8HN. ISBN 0 412 46610 4.

jute as a 'new crop' for fibre was deliberate by the first half of the nineteenth century. The early references to the jute plant up to the middle of the eighteenth century are concerned with identifying the pot-herb which Pliny (AD79) described as being used by the Egyptians (Watt, 1908). Royle (1855) propounded that the ancient Greeks referred to a pot-herb as 'Khorkhorus' from which the generic name *Corchorus* is derived. Indeed, *C. olitorius* is still grown as a leaf vegetable in Greece (Macedonia and Crete) and it is probable that *C. trilocularis* is the plant that Theophrastus records as being used in ancient Greece (Kavadas, 1956; Cheldriah and Miliarakis, 1980) and it was known to be used in Spain during the last century. *C. olitorius* is commonly grown as a vegetable crop and is marketed locally throughout the Middle East and in many parts of Africa, S.E. Asia and Latin America. The species *C. aestuans* L., *C. depressus* L., C. Christensen, *C. fascicularis* Lam., *C. tridens* L. and *C. trilocularis* L. are examples of edible greens which are gathered from the wild or collected as 'weeds' from the fields. They sometimes seem to occur cultivated. The leaves are used in sauces, relishes, green salads, soups and mucilaginous dishes either alone or mixed with other local leaf vegetables. The leaves are also consumed for various medicinal purposes in specific parts of Africa and Asia, although none are listed for Southeast Asia (Lemmens, *et al.*, 1989).

Distribution of the species listed above and their habitats and soil types are well described by Edmonds (1990).

5.2 MORPHOLOGY, ANATOMY AND PHYSIOLOGY OF GROWTH AND DEVELOPMENT

The two cultivated species of *Corchorus* have been well described in terms of their bast fibre cultivated forms in reviews by Kundu, (1956); Kundu *et al.*, (1959); Kirby, (1963); Dempsey, (1975) and Ghosh, (1983). The primary differences in leaf vegetable forms are the shorter, more robust plant stature, higher degree of axillary branching (in some cases) and leaf characters such as size, texture, pubescence and palatability. However, in the jute (fibre) producing countries of Asia young leaves of the fibre crop are mainly used for food, whereas in tropical Africa special vegetable forms with larger and more numerous leaves are preferred (IJO, 1987 and 1988).

Westphal-Stevels (1986 and pers. comm. 1990) conducted an agro-botanical survey of local vegetables in Cameroon and identified numerous vegetable types of *C. olitorius* varying in height, stem colour, pubescence, leaf and fruit shape and leaf production. Based on West African material *C. olitorius* was divided into two infra-specific categories: cultivar group *olitorius* including all vegetable types and cultivar group *textilis* including all cultivars used as a jute fibre crop. The cultivar group *olitorius* was further subdivided according to leaf shape and other characters:

Cv. group *olitorius*: leaf blade narrowly ovate to ovate, 4–13 × 2–5cm, margin serrate to serrate crenate.

Cv. *incisifolius*: leaf blade ovate to broadly ovate in outline, 4–7 × 2–5cm, 3–7 lobed or laciniate, margin deeply and coarsely serrate.

Cv. Géant de Bertoua: leaf blade ovate to broadly ovate, 9–16 × 6–13cm, margin broadly crenate, crenatures often spiculate.

A botanical description of *C. olitorius* (the most commonly used vegetable species) is presented below:

> Herbaceous annual with straight slender stem, 3–5 m height (fibre types), 30 cm–1.5 m height (vegetable types), branching near top. Leaves alternate, glabrous, light green; stipules two, linear, up to 1 cm long, deciduous; petiole 1–2 cm long; lamina lanceolate, 5–12 × 2–5 cm (larger and broader in some vegetable types). Flowers solitary or in few-flowered cymes, opposite leaves; sepals usually five, free, narrow, 7–8 mm long; petals usually five, deep yellow, 7–8 mm long; short corona separates petals from insertion of stamens; stamens 10 to many; free, filaments short, anthers small, bilobed; ovary superior, 5-locular with numerous ovules, style short with flattened stigma. Fruit a long cylindrical, 10-ridged, beaked capsule, 5–10 × 0.5–0.8 cm, dehiscing by five valves and with transverse septa between seeds. Seeds pyramidal, 1–2 mm long, dark greyish-blue in colour, 500 per g.

Early classical work by Burkill and Finlow (1907) on morphological and cultural characteristics identified five races of *olitorius*, three of which were grown as vegetables. Later work in India indicated at least eight forms of *olitorius* (excluding wild types) based on the colour of stem, petiole, leaf and pod variants. The pattern of anthocyanin pigmentation has been used as a key to the identification of cultivars and strains, vegetable types being generally less pigmented.

The cultivated species of *Corchorus* are typical short-day plants. In all photoperiods between 10–12 h, flowering is induced in about 30 days; the critical photoperiod appears to be about 12.5 h, above which flowering, and thus branching, is retarded (a desirable trait in fibre-producing cultivars).

C. capsularis withstands drought better at early stages of growth and is also more tolerant to flooding at mid and late growth stages. *C. olitorius* is more drought tolerant at later stages.

5.3 BIOCHEMICAL COMPOSITION

The leaves of *C. capsularis* contain a bitter glycoside, 'corchorin' and taste bitter on chewing; hence, it is often known in India and Bangladesh as 'Tita' (bitter) *pat*, whereas the almost tasteless leaves of *C. olitorius* are referred to

Table 5.1 Chemical composition

	Water (%)	Food energy[a]	% of dry matter: Nitrogen	Protein	Fat	Cellulose	Nitrogen free extract	Ash
C. tridens	81	308	3.77	23.59	3.35	8.41	53.85	10.80
C. trilocularis	78	318	4.86	30.36	4.81	7.86	49.32	7.65

[a]calories per 100 g dry matter.

Table 5.2 Mineral content (mg/g dry matter)

	Calcium	*Phosphorus*	*Magnesium*	*Sodium*	*Potassium*
C. tridens	1912	308	386	14	198
C. trilocularis	1428	292	417	135	270

as 'Mitha' (sweet) *pat*. There are, however, some non-bitter types of *C. capsularis*. Both forms develop a mucilaginous texture upon cooking.

The leaves are an important source of beta-carotene, the precursor of vitamin A. Jute leaves may contain as much as 18–22% protein; the bitter types of *C. capsularis* contain more than *C. olitorius* sweet types.

Detailed information on the biochemical analysis of cultivated jute leaves is lacking. However, Oliviera and Carvalho (1975) conducted a chemical survey of some edible leaves used in the local diet in Mozambique to assess their value as protein supplements in common diets. The two wild species *C. tridens* and *C. trilocularis* were included. The results of the analyses are presented in Tables 5.1–5.5.

Table 5.3 Amino acids pattern (mg/g dry matter)

	C. tridens	*C. trilocularis*
Arginine	11.98	16.73
Histidine	5.79	8.10
Lysine	11.65	12.85
Methionine	2.99	3.04
Cystine	1.56	1.62
Phenylalanine	11.71	11.90
Tyrosine	8.01	8.61
Leucine	23.74	22.68
Isoleucine	14.16	11.16
Valine	13.16	13.38
Threonine	8.38	9.89
Tryptophan	2.25	2.87

Table 5.4 Some indices of nutritional value of leaf proteins

	C. tridens	*C. trilocularis*
Chemical score (Mitchell and Block)	33	26
Limiting amino acid	(Met+Cys)	(Met+Cys)
Protein score (FAO/WHO)	35	35
Kuhnau's integrated index		
'Total value'	88	68
'Pure value'	83	68
'Supplementary value'	5	0
Modified essential amino acid index (Mitchell)	70	57
Biological value	50	30

In terms of food energy values the *Corchorus* species were superior to those of the species in the other genera assayed (various species of *Amaranthus, Portulaca, Momordica, Colocasia* and *Bidens*).

Studies on cultivated jute have shown that from 35–65% of total N, P, K and Ca removed from the soil by the crop is retained in the active green leaves.

5.4 AGRONOMY, YIELD AND PRODUCTION

The main areas of jute (fibre crop) cultivation lie several degrees north and south of the Tropic of Cancer, latitude 23°30′N, and between longitudes 86 and 92°E. In this area the crop responds favourably to the high relative humidity that averages 70–80%. The annual rainfall is 1500 mm or more with at least 250 mm occurring during each of the months of March, April and May. The range of mean temperature is 18–33°C. The area outside the main 'jute belt' where *Corchorus* is used primarily for vegetable purposes all possess the same general range of temperature, relative humidity and rainfall during the crop season (Dempsey, 1975).

Table 5.5 Niacin content

	Niacin	
	mg/100 g dry matter	*mg/16 g nitrogen*
C. tridens	5.31	22.50
C. trilocularis	7.00	23.00

In general, *Corchorus* occurs in lowland sites, but cultivated forms of *C. olitorius* were encountered at elevations of 2000 m and above in East Africa whereas the 'weedy' types were restricted to below 1400 m (IJO 1987 and 1988).

In the jute fibre crop belt young leaves of the fibre crop cultivars are consumed as vegetables. In most of the production areas overplanting is practised to secure uniform stands in the field of a fibre crop. After 30–45 days from sowing the excess plants are thinned out and their leaves used as vegetables (Hackett and Carolane, 1989). Elsewhere jute is exclusively grown as a vegetable. In such cases mostly branching and bushy plant types produce profuse fleshy leaves. The leaves are picked several times during the growing phase and some plants are allowed to flower and set seed to produce planting material for the following crop season. Vegetable types rarely attain a height of more than 1.5 m. They are often cultivated in small plots or kitchen gardens in admixtures with other vegetables, e.g. in East and West Africa variously with edible *Crotalaria* species, *Celosia argentea*, *Amaranthus hybridus* or *Gynandropsis gynandra*.

The green leaf yield of jute in Bangladesh and India is about 5 t/ha, in eastern China and Taiwan–China it is higher but in Brazil yields are much lower (1.5–2.0 t/ha). The average dry matter yield is 1600 kg/ha. No available statistics occur regarding production, *Corchorus* being a minor vegetable crop cultivated mostly for domestic consumption or local markets and rarely entering international trade. The crop is particularly popular in lowland West Africa and in the Middle East, more especially in Egypt from where dried, frozen jute leaves have been exported to USA to meet culinary demands by emigrants from the Middle East.

5.5 GENETICS AND BREEDING

Little attention has so far been given to the genetic improvement of *Corchorus* as a leaf vegetable. Nonetheless, Grubben (1977) in a detailed study of tropical vegetables accorded the highest priority to further research on *Corchorus* vegetables and ranked them with *Amaranthus* in terms of potential importance for nutrition.

The two cultivated (fibre) species are diploid ($2n = 14$) and both usually exhibit regular meiotic behaviour. Of the known wild taxa, the chromosome numbers of around 21 species have so far been reported. The majority are also diploid, but a number of tetraploid ($2n = 28$) species have now been reported. There is no cytological, or even morphological, information available on the remaining wild *Corchorus* species. Moreover, little work has been done on the cytotaxonomic relations of the wild species, either to each other, or to the cultivated species.

Both cultivated species are principally self-pollinated, although there is some crossing within each species. The range of crossing and breeding behaviour in jute has been reviewed by Basak (1988). Studies on the genetics of the bitter taste in *C. capsularis* have established a pattern of monogenic inheritance *Tn–tb* (bitter;non-bitter). The gene for bitterness was found to be linked with Br, the factor for branching with a crossing over percentage of 22.2, while there was no evidence of linkage with the allelomorphic series of the anthocyanin genes (Ghose *et al.*, 1948).

Despite the lack of serious attempts at breeding and varietal improvement, small collections of leaf vegetable *Corchorus* are maintained in Egypt, Libya, Sudan and Zimbabwe with duplicates conserved in Italy, Germany and India (Toll and van Sloten, 1982). A more recent large collection (about 2500 accessions) of *Corchorus* and allied bast fibre germplasm has been established by the International Jute Organisation and is conserved in Dhaka, Bangladesh with duplication in Australia. In view of the importance of leaf vegetable jute in the southern states of Nigeria, the National Horticultural Research Institute at Ibadan is active in collecting and evaluating material in its crop improvement programme. Local germplasm is assessed for plant, stem, leaf, fruit and root characters and resistance to root-knot nematodes. No efforts have been made to collect old introductions in many other parts of the tropics which have been known to have been used as leafy vegetables. It is not at all certain that the genepool of vegetable jute has been adequately explored or conserved.

5.6 FUTURE POTENTIAL

In general the consumption of leafy vegetables in the tropics is far from optimal, particularly considering that the vitamin A requirements in such regions frequently depend on vegetable products, and especially on leaf sources (Grubben, 1977; van Sloten, 1984).

Jute leaves are one of the most important local market vegetables in the tropics. The cultivated forms thrive in garden plots with the minimum of cultural attention and inputs. Many wild species and 'weedy' types exist in and around homesteads and are also used as food plants. Jute leaves are popular for the variety of culinary preparations in which they are a favoured ingredient, especially for the mucilaginous texture which they impart to local dishes in Africa and Asia. The leaves are used fresh or dried and can be stored after drying and used later during periods of scarcity.

The importance of wild species of *Corchorus* as vegetables and salad plants in more than half the countries in Africa should encourage an investigation of their similar usage in Asian countries. Significant economic and nutritional contributions within Asia might result from such a study (Edmonds, 1990). A greater use of these leaves will be an effective and inexpensive method of improving the quality of carbohydrate-based diets (maize, rice, etc.) without a drastic change in food habits, since the plants are already pantropical in distribution. IBPGR recognizes the need for conserving and investigating the wider potential of minor leaf vegetables. In the case of *Corchorus* the Nigerian national programme (Nath and Denton, 1979, 1980) and national programmes in Southeast Asia (IBPGR, 1980) might consider a collaborative endeavour to introduce more of the leafy forms into Asia.

At the same time, the wide use of *Corchorus* (leaves and other plant parts) for medicinal purposes might be worthy of future investigation. Properties including laxative and antihelminthic (extracted from dried leaves – Myser Ali, pers. comm.) are locally popular in many tropical countries.

REFERENCES

Basak, S.L. (1988) Breeding behaviour and isolation distance in jute and kenaf – a review. International Jute Organisation, Dhaka, Bangladesh.

Burkill, I.H. and Finlow, R.S. (1907) Races of jute. *Agric. Ledger*, **6**, 41–137.

Cheldriah, Th. and Miliarakis, S. (1980) [*Lexicon of Local Names of the Greek Flora*] (in Greek), Athens.

Dempsey, J.M. (1975) *Fibre Crops*, University of Florida Press, Gainsville.

Edmonds, J.M. (1990) *Herbarium survey of African* Corchorus *L. species*, Systematic and Ecogeographic Studies on crop Genepools 4. IBPGR/IJO, Rome, Italy, 284 pp.

Ghose, R.L.M., Rao, K.R. and Kundu, B.C. (1948) The genetics of *Corchorus* (Jute) V. The inheritance and linkage relations of bitter taste, anther and corolla colour. *J. Genet.*, **49**, 14–22.

Ghosh, T. (1983) *Handbook on Jute*, FAO Plant Production and Protection Paper No. 51, FAO, Rome, Italy.

Grubben, G.J.H. (1977) *Tropical Vegetables and their Genetic Resources*, (eds H.D. Tindall and J.T. Williams), IBPGR, Rome, Italy.

Hackett, C. and Carolane, J. (eds.) (1989) *Edible Horticultural Crops. A Compendium of Information on Fruit, Vegetable, Spice and Nut Species*, Parts I & IV, Academic Press, Sydney, Australia.

IBPGR (1980) *A Co-operative Regional Programme in Southeast Asia*, Report of the third regional IBPGR committee meeting, IBPGR, Rome, Italy.

IJO (1987 and 1988) Germplasm Collection in East Africa. Reports on IJO germplasm collecting missions to Kenya and Tanzania. (ed. I.R. Denton), International Jute Organisation, Dhaka, Bangladesh.

Kavadas, D.S. (1956) *Illustrated Botanical and Phytological Dictionary* (in Greek), Athens, Greece.

Kirby, R.H. (1963) *Vegetable Fibres*, Leonard Hill, London.

Kundu, B.C. (1951) Origin of Jute. *Ind. J. Genet. and Pl. Breed.*, **11**, 95–9.

Kundu, B.C. (1956) Jute – World's foremost bast fibre. I. Botany, agronomy, diseases and pests. *Econ. Bot.*, **10**, 103–33.

Kundu, B.C., Basak, K.C. and Sarcar, P.B. (1959) *Jute in India*, Indian Central Jute Committee, Calcutta, India.

Lemmens, R.H.M.J., Jansen, P.C.M., Siemonsma, J.S. and Stavast, F.M. (eds) (1989) *Plant Resources of South-East Asia. Basic List of Species and Commodity Grouping*, PROSEA Project, Wageningen, Netherlands.

Nath, P. and Denton, L. (1979) Vegetable germplasm in Nigeria. *Plant Gen. Resour. Newslett.*, **39**, 24–5.

Natu, P. and Denton, L. (1980) Leaf vegetables in Nigeria. *Plant Gen. Resour. Newslett.*, **42**, 14–17.

Oliviera, J.S. and Carvalho, M.F. (1975) Nutritional value of some edible leaves used in Mozambique. *Econ. Bot.*, **29**, 255–63.

Purseglove, J.W. (1968) *Tropical Crops: Dicotyledons*, Longman, London.

Royle, J.F. (1855) *Fibrous Plants of India*, Smith, Elder and Co., London.

Sloten, D.H. van (1984) The genetic resources of leafy vegetables, in *Crop Genetic Resources: Conservation and Evaluation*, (eds J.H.W. Holden and J.T. Williams), Allen and Unwin, London, pp. 240–8.

Toll, J. and van Sloten, D.H. (1982) *Directory of Germplasm Collections 4. Vegetables*, IBPGR, Rome.

Watt, G. (1908) *Commercial Products of India*, John Murray, London, pp. 405–27.

Westphal-Stevels, J.M.C. (1986) Local vegetables in Cameroon: *Corchorus* species used as a vegetable. *Acta Horticulturae*, **182**, 423–5.

6

Vegetable cactus (*Opuntia*)

E. Pimienta-Barrios

6.1 TAXONOMY, HISTORY AND ECOLOGY

The genus *Opuntia* has been divided into subgenera with subgenus *Cylindropuntia* containing the chollas or species with cylindrical stems and *Platyopuntia* containing species with flattened leaf-like stems (botanically cladodes) or 'cactus pads'. Bravo (1978) expanded this division and the species of interest for vegetables are now in subgenus *Opuntia*.

Whereas in the centre of diversity, Mexico, a large number of species are to be found, one *O. ficus-indica* (L.) Miller, the widely known prickly pear, was taken to N. Africa from Mexico in the sixteenth century, and is naturalized in the Mediterranean and many other parts of the world. The literature shows how this species has become important for fodder: it is cultivated in arid and semi-arid zones of N. Africa in well over 200 000 ha (le Houérou, 1979) and used for a variety of purposes, e.g. for fodder and soaked stems for soap making as in Malawi (Williamson, 1975) and the fruits have become part of the Mediterranean culture. Variety *inermis* is a spineless form of the species and has spread widely.

The semi-arid regions of the central part of Mexico are the centre of diversity and the use of prickly pear as a crop dates back to ancient mesoamerican civilizations. *O. ficus-indica* is not the only edible fruit-bearing species: *O. streptacantha* Lemaire, *O. lindheimeri* Engel and *O. robusta* Wendland are also important. *O. ficus-indica* has hybridized with *O. streptacantha* and with other species of the genus distributed northwards into the southwest of the US (Benson, 1969). Martin *et al.* (1987) list an additional eleven species gathered for fruits in various parts of tropical America, some locally distributed in Mexico and the USA, and there appears to be a secondary centre of diversity in Ecuador. Apart from *O. ficus-indica*, one other species has been more widely distributed: *O. tuna* Mill., to Hawaii, Micronesia and Japan.

Pulses and Vegetables. Edited by J. T. Williams. Published by Chapman & Hall, 2–6 Boundary Row, London SE1 8HN. ISBN 0 412 46610 4.

The species is associated with a host of vernacular names. Prickly pear is also called Indian fig, and in the New World, nopal, cardon de Mexico, chumbo, higo or tuna and in French, chardon d'Inde, fiquier d'Inde. The cactus pads are called nopalitos (the name tuna refers to fruits) and are used as a vegetable. One other species, *O. fulgida* Engel, the jumping cholla or velas de coyote, is used as a minor vegetable by the Seri Indians when fruits are gathered green and used fresh or cooked (Felger, 1979).

Gathering of cactus pads has been traditional in the centre of diversity and the vegetable forms an important part of the Mexican vegetable market and traditional cuisine. The most important wild species that are used as a source of nopalitos are *O. robusta*, *O. streptacantha*, *O. leucotricha* DC, *O. hyptiacantha* Weber, and *O. chavena* Griffiths (Figueroa, 1984) and the most commonly used in plantations is spineless forms of *O. ficus-indica*.

The origins of the Mexican culture are linked to the prickly pear or nopal. One of the most important pre-Hispanic cultures, the Aztecs, founded their city where they found the sign which their God had given to them: an eagle on a prickly pear, devouring a serpent.

The Indian tribes that inhabit the semiarid regions always used *Opuntia* plants as a food source in times of scarcity. It is well documented by the Indian codex, that numerous wandering tribes agree during the spring and summer time to gather young cactus pads and fresh fruits from the wild *Opuntia* populations. The pads were prepared in different ways: the young tender pads were cooked, roasted over flat earthenware ('comal') and dressed with meat of wild animals (deer, armadillo, or rabbit) and with wild vegetables (onion and tomato) and with diverse seeds (pumpkin, peanut, or cacao) (Sodi, 1968; Martin del Campo, 1957; Sanchez-Mejorada, 1982).

It is thought that the wild *Opuntia* populations influenced the settlement of important pre-Hispanic civilizations in the plateaus of the semiarid regions of the central part of Mexico, because of dependence on prickly pear as a source of food and also as a medicinal plant (Hernandez, 1985).

The customs of use of prickly pear were later adopted by the Spanish explorers, conquerors and settlers who followed Columbus. Spanish chroniclers of the conquest made frequent reference to the use of prickly pear as a human food and also as a medicinal plant (Bravo, 1978).

Species of *Opuntia* are adapted to the arid and semiarid vegetation of Mexico. They have a wide range of low temperature sensitivities (Uphof, 1916) and cultivation is often limited by low wintertime temperatures (Russell and Felker, 1987b). A great deal of information on the ecological tolerances can be derived from the areas of commercial plantations in Mexico (Figure 6.1). The most important areas devoted to the production of fresh fruits and young pads are in the highlands of the semiarid regions of the central-north part that comprises the states of Zacatecas, San Luis Potosi, Aguascalientes, Jalisco, and Guanajuato in which the cultivated area

Figure 6.1 Prickly pear commercial plantation devoted to the production of nopalitos.

is estimated to be greater than 35 000 ha. The other centre is situated in the central–south region which includes territorial parts of the states of Hidalgo, Mexico, Tlaxcala and Puebla, in which prickly pear grows in an area of approximately 15 000 ha. There are other production centres, but they are dispersed and occupy relatively small areas.

The commercial plantations in the central–north region are found at elevations from 1800 to 2200 m. The summer rainfall varies from 400 to 500 mm per annum with a bimodal pattern of distribution. Some years rains occur during the winter months. However, in most of the years the dry period coincides with the winter and spring seasons. The average temperature during the year varies from 16 to 18°C. Most of the soils are shallow and derived from igneous rocks, with a sandy-loam soil texture. The soil pH ranges from moderately acid (5.1–6.1) to slightly acid (6.1–6.7). The organic matter content is low (0.8–1.8%). The natural vegetation is composed mainly of grasses of the genus *Buchloe, Sporobolus, Bouteloua,* and *Muhlenbergia.*

In the central–south part, the cultivated prickly pear populations are grown on different types of soils. The main types are: feozem, vertisol, and luvisol. The feozem is the most abundant type, which is characterized by the presence of a superficial dark layer, rich in organic matter and mineral nutrients. The luvisols are distinguished by a subsoil clay accumulation, with a pH that fluctuates from slightly acid to extremely acid. The vertisols are more fertile than the feozem and luvisols with a high clay content.

In this region the commercial plantations occur in areas with an elevation from 1800 to 2400 m and a rainfall from 400 to 700 mm per annum. The annual average temperature varies from 14 to 18°C.

Prickly pear plantations that are established in localities where the mean annual rainfall is less than 300 mm are severely affected by drought. The effects of drought are further exacerbated by browse from wildlife that face a scarcity of food from natural vegetation. Another important limiting factor is soil pH, since few commercial varieties grow satisfactorily in alkaline soils; most prefer acid soils – although *O. ficus-indica* thrives on calcareous soils.

6.1.1. **Research interest**

Opuntia was not singled out as an underexploited tropical crop by the Board on Science and Technology for International Development of the US National Academy of Sciences in the 1970s. Nor did the comprehensive review of tropical vegetables and their germplasm by the International Board for Plant Genetic Resources (Grubben, 1977) accord any priority.

A resurgence of interest by Mexican universities from the mid-1970s to

Figure 6.2 Packing of young fresh cladodes.

the present; expansion of export of cactus pads to the United States, fresh, canned, processed or frozen, and enhanced interest in the value of *Opuntia* in semi-arid agroforestry (Felker, 1988) and for commercial production and enhanced use for fodder in the southwest United States has led to a substantial research base, since the time of the NAS and IBPGR studies. The Texas Prickly Pear Council was established in 1989 to further education, research, production, utilization and marketing of *Opuntia*. Opportunities exist for technology transfer to other arid and semiarid areas and for the development of new cultivars to replace old introductions. Justification for research attention and status reports are provided in Borrego (1986), Flores and Aguirre (1979), Pimienta (1990) and Russell and Felker (1987a).

Processing is well established. In Mexico, most wild material is supplied to packing companies located in the city of San Luis Potosi. These companies process from 2000 to 3000 tons of nopalitos each year although annual production from cultivation is over 232 000 tons from more than 4000 ha and is dealt with by other companies (Figure 6.2).

The pads are bottled whole or chopped, in different manners; in water and iodized salt and with vinegar and combined with other vegetables, e.g. onion, carrot. The packing companies commercialize their products in local markets and also export to United States markets. There are current research interests in value-added products (Sanchez *et al.*, 1990) and marketing requires further research.

6.2 VARIATION AND DOMESTICATION

The variability of wild and domesticated prickly pears has been well studied. Several authors believe that the variability arose via natural hybridization associated with polyploidy and geographical isolation (Gibson and Nobel, 1986; Pimienta and Mauricio, 1989). The differences in fruit size found in wild and cultivated prickly pear populations are undoubtedly due to differences in ploidy levels, since previous cytogenetic studies revealed the existence of different levels of ploidy (Heras *et al.*, 1988; Brutsch, 1984). An interesting conclusion that emerged from these studies is that the prickly pear varieties and forms with highest chromosome numbers are commonly found in cultivated conditions ($2n = 66$ and $2n = 88$). In contrast, the lower chromosome numbers are found in wild populations ($2n = 22$ and $2n = 44$).

Changes in ploidy levels are commonly expressed by an increase in vegetative vigour (cladode or pad size) and reproductive vigour (fruit size) and probably played an important role in domestication. During the early stages of domestication, phenotypes with higher ploidy levels and vigorous cladodes and bigger fruits were probably preferentially selected and established vegetatively in backyards. Over time the rural populations dispersed in the semiarid regions of Mexico, and not only became reservoirs of prickly pear germplasm, but also contributed to the evolution of the species.

The artificial sympatric conditions created by the number and diversity of species that grow close together in small areas and the opportunities for geneflow greatly stimulated the evolution of cultivars in backyard environments. Thus, backyards constitute a major genepool of germplasm that is invaluable for future breeding and is a unique biological material for basic evolutionary studies.

While there are studies on prickly pear taxonomy, there is a lot of confusion. The main reason for this confusion is because the majority of the descriptions were made with small samples, without considering the variability that exist in their natural habitats. Some of the classified specimens are in reality varieties, geographic forms or hybrids (Bravo, 1978). There is an urgent need for modern biosystematic research in prickly pear, in order to establish a more coherent classification of both wild species and cultivars.

The first commercial plantations were established using vegetative material from outstanding phenotypes from the backyards of rural homes. The selection of these phenotypes must have been quite rigorous, since the prickly pear growers needed to consider production potential, adaptation and resistance to environmental stresses, predation caused by wildlife and aspects related to market demands.

The most important commercial plantations are located in the highlands of the locality of Milpa Alta, close to Mexico city, this being one of the most important consumption centres in Mexico. Milpa Alta is an outstanding example of the use of steep land for agricultural practices. This zone was originally deforested for the cultivation of staple crops such as maize, with poor results, since their productivity was dependent on the highly variable and erratic rainfall. Thus, the change from maize to prickly pear allowed the transformation of marginal land into productive land.

Other important production centres are located in tropical and subtropical climates in the states of Jalisco, Colima and San Luis Potosi. The absence of low temperatures during the winter season allow those commercial plantations to produce young cactus pads during the winter season, a time of the year in which the plantations located in more temperate climates have reduced production due to the low temperatures.

6.3 NUTRITIVE VALUE

The nopalitos are constituted principally of water (92%), carbohydrates (4–6%), proteins (1–2%), minerals (1%) and a moderate content of vitamins, mainly A and C. Due to the fact that prickly pear is a CAM (Crassulacean acid metabolism) plant, the organic acid content of the young pads varies during the day; the pH ranges from 5.2 at evening to 4.4 at early morning,

Table 6.1 Nutritive value of nopalitos compared to those of lettuce and spinach

Chemical component	*Nopalito*	*Lettuce*	*Spinach*
Water (%)	95.0	95.5	90.7
pH	4.6	5.9	6.4
Protein (%)	1.5	0.85	3.2
Fat (%)	0.4	0.09	0.3
Fibre (%)	1.54	–	–
Total carbohydrates (%)	5.0	1.8	4.3
Total chlorophyll (μg/g)	15.5	–	490.0
Vitamin C (mg/100 g)	22.0	15.0	51.0[a]
β-carotene (μg/100 g)	31.0	19.0	55.0
Niacin (mg/100 g)	0.46	0.28	0.6
Thiamin (mg/100 g)	0.14	0.06	0.1
Riboflavin (mg/100 g)	0.6	0.06	0.2
Calcium (mg/100 g)	110.0	19.0	93.0[b]
Phosphorous (mg/100 g)	20.0	21.0	51.0
Iron (mg/100 g)	1.95	0.5	3.0

[a]86.0 and [b]62.0 in FAO Food Composition Tables.
Sources: Arthey (1975); Neri (1991); Cantwell (1991); and data from Instituto Nacional de Nutricion, Universidad de Guanajuato, and Comision Nacional de Fruticultura, SARH.

and the percentage of malic acid varies from > 0.5% at 8 00 a.m. and <0.1% at 4.00 p.m. (Cantwell, 1991; Neri, 1991).

Table 6.1 shows a comparison of the nutritive values of nopalitos and two other important leafy vegetables, lettuce and spinach.

The nutritive values of nopalitos compare favourably with the other two vegetables. The nopalitos may be considered a good source of vitamins and minerals, mainly of vitamin C and calcium. The vitamin C content is superior to the values reported for French beans, lettuce, carrot and close to the values registered for spinach beet, amaranth, chayote, asparagus, pea and tomato. However, among vegetables it may be considered as one with relatively low vitamin C content. In broccoli and horseradish, the values of this vitamin may be as high as 120 mg/100 g of fresh tissue (Arthey, 1975).

Although carbohydrate content is relatively low in comparison with other vegetables, it is almost double that reported for lettuce and slightly superior to spinach. The protein of the nopalitos is superior to the values reported for lettuce and about 50% of the values of spinach. Recent screening of clones promises much higher protein content; one has been identified with over 10% (Felker, pers. comm.).

In most of the vegetables, starch is the main carbohydrate reserve. However, in prickly pear, in addition to starch, mucilage functions also as a carbohydrate reserve (Sutton *et al.*, 1981). Recent work has demonstrated the importance of mucilage in the control of lipoprotein levels and cholesterol (Fernandez *et al.*, 1990).

Another important feature common to most cactuses is the relatively high levels of calcium. These high levels of calcium in nopalitos may represent the accumulation of calcium oxalate (Nobel, 1983).

6.4 AGRONOMY, YIELD AND PRODUCTION

Since prickly pears are CAM plants, they have high water-use efficiency and are drought tolerant. They are four to five times more efficient than grasses in converting water to dry matter, so prickly pears are well suited to agricultural systems with water shortage (Russell and Felker, 1985; Sutton *et al.*, 1981).

Commercial plantations devoted to the production of nopalitos are normally established using population densities that vary from 25 000 to 40 000 plants per ha, planted about 30–40 cm apart within rows and 80–100 cm between rows. Planting is usually through the use of mature cladodes (2–3 years old) and planting time is between March and May, after the cold season and before the rain starts.

Production starts relatively soon, since the first crop can be harvested in 2–3 months. The yields are relatively variable at between 30 and 80 tons/ha per year, although some commercial plantations have reached yields close to 100 tons/ha. The cladodes are harvested when young and tender and are harvested all through the year, although productivity is higher during the spring season. At this time of the year, the growers harvest pads two and even three times per week.

During the summer season, the yields show slight decreases, mainly due to the start of the rainy season that lowers the temperature. The harvesting frequency at this time is reduced to one harvest per week. In the spring season, the pads take about seven days to reach the harvest size, whereas during the summer, the harvest size is reached in *ca.* 15 days. The time of harvesting is determined by the size of the cladodes; normally they are harvested at about 20–25 cm length and the weight is about 90–100 g. Older cladodes start lignification. Cladode growth is drastically reduced at the middle of autumn and all through the winter.

The productivity of a commercial plantation depends on the plantation age. The yields are low during the first two years (3–5 ton/ha) but after the third year show a significant increase to > 40 tons/ha. These relatively high yields are maintained until the plantation is about ten years old, after which time the yields show progressive reduction. The average time in which a plantation remains productive is about 15 years.

In Mexico and the south of Texas, intensive production systems have been developed using microtunnels (see Figure 6.3) to assure high yields and production of young cladodes during the coldest months of the year,

Figure 6.3 Intensive production of 'nopalitos' using microtunnels.

providing higher prices, when produce is normally scarce in both cultivated and wild populations due to low temperatures. Under these intensive protected production systems, it is feasible to obtain 30 kg/m^2. In the microtunnel, the cladodes are planted 10 cm apart between plants and 25 cm between rows (Grajeda, 1977).

In the commercial plantations devoted to vegetable production, the plants are not allowed to reach more than 1.5 m in height, since larger plants are difficult to harvest, and they tend to produce more fruits than young pads.

Growers normally apply high levels of organic matter (above 100 ton/ha of bovine manure), and in a few cases apply mineral fertilizers. Experimental work has shown a positive response of prickly pear to the combined application of both manure and mineral fertilizers, even in the absence of irrigation. Yields close to 200 ton/ha were obtained with 200 kg/ha of nitrogen and 100 tons of bovine manure without irrigation.

The application of low rates of nitrogen increases the percentage of crude protein significantly and high nitrogen treatment (224 kg/ha) is needed to meet requirements for fodder for lactating cows. Probably application of nitrogen every 2 years at this rate would be adequate for fodder for both dry and lactating cows (Gonzalez and Everitt, 1990). Application of phosphorus (112 kg/ha) doubles P content which is normally low in prickly pear.

Figure 6.4 Pruning is a common practice in commercial plantations, because it increases yield and facilitates harvest of 'nopalitos'.

With regard to water supply, most of the commercial plantations are in localities where the summer rainfalls are between 600 and 1000 mm, so irrigation is practised only during the spring dry season and in smaller volumes three to four times a year. Each irrigation applies the equivalent of 10 mm of rainfall. Once the rains start, there is no need for irrigation. Sprinkler irrigation is one of the main irrigation systems used.

Most of the commercial Mexican plantations do not apply water, especially those in localities where the summer rainfall is over 800 mm per annum.

Empirical observations on commercial plantations show that pruning is effective in promoting pad yields, especially during the spring season (Borbolla and Ayala, 1987); see Figure 6.4.

Detailed reports analysing the material costs and labour of prickly pear plantations requirements are available (Montalvo *et al.*, 1990).

During recent years, prickly pear growers have become aware of increased damage caused by pests and diseases that in particular severely affect fruits. Research oriented to the control of pests and diseases on prickly pear is scant. It is paradoxical, that there is more information on the biological control of natural prickly pear populations using phytophagous insects, than in the control of those insects. In spite of lack of information, growers have started applying agrochemicals to reduce insect damage. This situation needs to be prevented, to avoid environmental problems.

The most important diseases are named 'swelling of cladodes' and 'bud proliferation'. The first disease is caused by a mycoplasma and the second one by a spiroplasma (Pimienta, 1990). Among the pests the most important are different types of corkborers, e.g. zebra worm *Olycella nephelasa*, white worm *Lanifera cycladea*, and phytophagous insects that cause damage in the areoles, e.g. *Cylindrocopturus biradiatus* and tissue yellowing, e.g. cochinilla or grana *Dactylopius* spp.

6.4.1 **Post-harvest physiology**

The young cladodes have a relatively short storage life (10 days) like other vegetables and perish rapidly in storage at ambient conditions. However, storage at 10°C and 80–85% RH allows young tender pads to remain in excellent condition for about 30 days. Those stored at relatively low temperatures (< 10°C) have chilling injury (Ramayo *et al.*, 1978a) and also quality changes (Ramayo *et al.*, 1978a; Cantwell, 1991). Storage at 5°C causes increased tissue acidity, and storage at higher temperatures, 20°C, causes decreased tissue acidity (Cantwell, 1991).

The nutritional value of the pads changes after harvest (Table 6.2). The most important changes are: decrease in fresh weight, reduction in the content of both total and reducing sugars, and an increase in pH and protein content.

6.4.2 **Mechanical damage**

Any mechanical damage caused during harvest increases the possibility of microbial spoilage, even under optimal storage conditions. The principal organisms responsible for spoilage are moulds of *Penicillium*, *Aspergillus*, and *Alternaria* species. Careful harvesting by cutting at the base of insertion reduces microbial spoilage (Ramayo *et al.*, 1978b).

Table 6.2 Changes in some chemical components in young cladodes at different intervals of time after harvest under light conditions (Neri, 1991)

Days after harvest	*Fresh weight (g)*	*pH*	*Total sugars g/g fw*	*Reducing sugars g/g fw*	*Protein content g/g fw*
2	82.2	4.36	0.31	0.35	1580
8	70.8	4.78	0.07	0.12	1672
15	68.8	4.81	0.09	0.08	1559
22	64.5	4.98	0.11	0.05	2088

6.5 BREEDING

To the present, cultivars have been selected from the wide genepool on the basis of pads with characteristics desirable for human consumption. Consumers prefer two varieties: 'duraznillo' (*O. leucotricha*) and 'tapon' (*O. robusta*) for use in different kinds of dishes. Packing companies have specific demand for the variety 'tapon'. There are different reasons for this preference; among the most important are:

1. their round shape and green-clear colour with dispersed red spots that make the pads visually attractive;
2. the cuticle is easily peeled;
3. they are rapidly cooked and the pads do not disintegrate when they are cooked for bottling (the majority of prickly pear varieties do not withstand the cooking process); and
4. this variety releases less mucilage during cooking, in comparison to the other varieties (Figueroa, 1984).

An important volume of pads is still harvested from wild *Opuntia* populations. Selection of better cultivars and their use in plantations and backyards will help to safeguard the wild genepool.

Great opportunities exist for cultivar development through crossing and testing progeny both in the centre of diversity and in other arid areas of the world. There is no fully comprehensive genetic resources collection and conservation programme although a proposal was developed in the late 1980s. Maybe other uses, for instance for medicinal purposes, will add weight to the need for such a programme.

6.6 FUTURE POTENTIAL

Since processing and export markets are already in place for Mexico and Southern California (and expanding in Chile), further developments will depend on enhanced marketing. One innovative marketing enterprise in the USA has found that alternate names – cactus pear for tuna (fruits) or cactus leaves for nopalitos (cactus pads) – help the purchaser. Almost certainly complementary products – jellies, jams, ice creams and desserts, emphasizing low-sodium, low-sugar products will help consumer awareness (Caplan, 1990; Sanchez *et al.*, 1990). Additionally, enhanced production for fodder can be expanded in numerous parts of the world and possibly as a spin-off of enhanced attention to cultivar development for vegetables.

During the summer months, an important part of the production is not sold on the vegetable market, since it is used for dairy cattle fodder. The local dairy operations supply the fresh manure which is applied to the cultivated prickly pear fields for moisture conservation as well as the main source of fertilizer.

The nopalitos are also used in the cosmetic industry for the production of different products such as shampoos, hair conditioners, and nopal hydrating and moisturizing facial creams. A company in the city of Guadalajara has developed and commercialized some of those products in both national and international markets with apparently satisfactory results. Most of the properties of the shampoos, conditioners and creams are attributed to the mucilage present in the pads.

The tender cladodes have been traditionally used since the pre-Columbian times by the Mexican populations as a medicinal plant. A common practice which survives to this day is the use of heated young cladodes as poultices for reducing swelling. The cladodes were also used to reduce burning sensations of kidneys and urine. The juice or distilled fluid is excellent for cases of malignant and bilious fever and helps to heal ulcers (Meyer and McLaughlin, 1981; Bravo, 1978).

Ethnobotanical studies in rural Mexico have revealed an extensive use of prickly pear as a treatment for the control of diabetes mellitus, by eating young pads before each meal (Morales, 1990). Reports of the use of prickly pear to treat diabetes come also from Australia and South Africa (Meyer and McLaughlin, 1981).

Despite the traditional evidence for medicinal use, it was not until the establishment of the Mexican Institute of Medicinal Plants in early 1970, that the hypoglycaemic properties of prickly pear cactus were studied in a scientific manner. Clinical trials using fresh pads before each meal for ten days, showed lowering of cholesterol, glucose and insulin levels, (Trejo Gonzalez *et al.*, 1991; Frati-Munari *et al.*, 1983). It is now being used for the treatment of hypoglycaemia, diabetes, high blood cholesterol and obesity.

Recently in an elegant work, Fernandez *et al.*, (1990), showed that a pectin (mucilage) isolated from prickly pear reduces low density lipoprotein and high density lipoprotein cholesterol levels.

ACKNOWLEDGEMENTS

The author thanks Dr J. Trevor Williams of the International Program for Tropical Tree Crops Conservation and Development for his valuable help in the manuscript edition. Grants from the Consejo Nacional de Ciencia y Tecnologia (D112–904 333) and the Departmento de Investigacion Cientifica y Superaction Academica de la Universidad de Guadalajara (89/MB/89–1225/01/3783), provided essential support for the studies cited as unpublished data from the author's laboratory.

REFERENCES

Arthey, D.V. (1975) *Quality of Horticultural Products*, Butterworths, London.
Benson, L. (1969) *The Native Cacti of California*, Stanford University Press, Stanford.

Borbolla, F.L. and Ayala, C., J.C. (1987) Estudio de factibilidad tecnica y economica para el establecimiento de una planta empacadora de nopal de verdura (*Opuntia ficus-indica* (L.) Miller) en Milpa Alta, D.F. Tesis Profesional, Universidad Autonoma Chapingo.

Borrego, E.F. (1986) *El Nopal*, Universidad Autonoma Agraria, Antonio Narro', Saltillo, Coahuila, Mexico.

Bravo, H.H. (1978) *Las Cactaceas de Mexico*, UNAM, segunda ed.

Brutsch, O.M. (1984) Prickly pear (*Opuntia ficus-indica*) cultivation in Southern Africa, *Symposium on Agricultural use of Cactaceae: Prospects and problems*, Congress of the International Organization for Succulent Plant Study, Frankfurt.

Cantwell, M. (1991) Calidad y fisiologia postcosecha de tunas y nopalitos, *Memorias des Simposium Internacional de Nopal y Tuna*, Lagos de Moreno, Jalisco, Mexico, Octobre, 1991.

Caplan, K. (1990) Marketing strategies for cactus pears and cactus leaves for the 1990s, in *Proceedings of the First Annual Texas Prickly Pear Council*, Kingsville, Texas, pp. 85–9.

Felger, R.S. (1979) Ancient crops for the twenty first century, in *New Agricultural Crops*, (ed G.A. Ritchie), AAAS/Westview Press, Boulder, Colorado, pp. 5–20.

Felker, P. (1988) Semiarid agroforestry-multiple values, in *Challenges in Dryland Agriculture; A Global Perspective*, Texas Agricultural Experiment Station, Bushland, Texas, pp. 762–3.

Fernandez, M.L., Trejo, A. and McNamara, D.J. (1990) Pectin isolated from prickly pear (*Opuntia* sp.) modifies low density lipoprotein metabolism in cholesterol-fed guinea pigs. *J. Nutr.*, **120**, 1283–90.

Figueroa, H.F. (1984) Estudios de las nopaleras cultivadas y silvestres sujetas a recoleccion para el mercado en el altiplano potosino-zacatecano. Tesis de Licenciatura, Escuela de Agronomia, Universidad Autonoma de San Luis Potosi, Mexico.

Flores, V.C.A. and R. Aguirre (1979) *El Nopal como Forraje*, Universidad Autonoma Chapingo.

Frati-Munari, A.C., Fernandez-Harp, J.A. Bañales-Han, M. and Ariza-Andraca, C.R. (1983) Disminucion de glucosa e insulina sanguineas por nopal (*Opuntia* sp.) *Arch. Invest. Med.*, **14** (3), 269–74.

Gibson, C.A. and Nobel, P.S. (1986) *The Cactus Primer*, Harvard University Press, Cambridge, MA.

Gonzalez, C.L. and Everitt, J.H. (1990) Fertilizer effect on the quality and production of prickly pear cactus and its wildlife value, in *Proceedings of the First Annual Texas Prickly Pear Council*, Kingsville, TX, pp. 3–13.

Grajeda, G.J. (1977) Producción de nopal para verdura en invierno. Informe Anual, Sección de Fruticultura, Colegio de Postgraduados, Chapingo, Mexico.

Grubben, G.J.H. (1977) *Tropical Vegetables and their Genetic Resources*, IBPGR, Rome.

Heras, J., Palomino, G. and Scheinvar, H.L. (1988) Estudios cariotipicos en tres especies del genero *Opuntia* (Cactaceae) del valle de Mexico. 3a Reunion nacional y la 1a Reunion Internacional. Universidad Autonoma Agraria Antonio Narro, Buenavista, Saltillo, Caohuila, pp. 109.

Hernandez-X, E. (1985) *Biologia Agricola*, CECSA, Mexico.

le Houérou, H.N. (1979) Resources and potential of the native flora for fodder and some passive production in the arid and semi-arid zones of North Africa, in *Arid Land Plant Resources*, (eds J.R. Goodwin and D.K. Northington), Texas Tech. University, Lubbock, TX, pp. 358–401.

Martin, F.W., Campbell, C.W. and Ruberte, R.M. (1987) *Perennial Edible Fruits of the Tropics*, USDA, Washington, DC.

Martin del Campo, R. (1957) Las cactaceas entre los mexica. *Cact. y Suc. Mex.* **11** (2), 27–38.

Meyer, N.B. and McLaughlin, J.L. (1981) Economic uses of *Opuntia*. *Cactus and Succulent Journal*, **53**, 107–12.

Montalvo, A., Montalvo, D., Daniels, L. and Felker, P. (1990) A preliminary economic assessment of growing cactus for fruit production in south Texas, in *Proceedings of the First Annual Texas Prickly Pear Council*, Kingsville, TX, pp. 33–7.

Morales, B.L. (1990) Uso y experiencia de la utilizacion del nopal (*Opuntia ficus-indica*) dentro del campo de la medicina, in *El Nopal su Conocimiento y Aprovechamiento* 3a. Reunion Nacional 1a. Internacional, (eds J.J. Lopez G. and M.J. Ayala O.), Universidad Autonoma Agraria 'Antonio Narro', Buenavista, Saltillo, Coah., Mexico, pp. 352–8.

Neri, L.C. (1991) Cambios fisiólogicos durante el proceso de senescencia en cladodios jovenes de nopal (*Opuntia ficus-indica* (L.) Miller), Tesis de Licenciatura, Universidad de Guadalajara.

Nobel, P.S. (1983) Nutrient levels in cacti in relation to nocturnal acid accumulation and growth. *American Journal of Botany*, **70** (8), 1244–53.

Pimienta, B.E. (1990) *El Nopal Tunero*, Serie Libros Tiempos Ciencia, Universidad de Guadalajara.

Pimienta, B.E. and Mauricio, R. (1989). Variacion en componentes del fruto maduro entre formas de nopal (*Opuntia* spp.) tunero. *Revis. Fitotec.* Mex., **12**, 183–96.

Ramayo, R.I., Saucedo, V., C. and Laksminarayana, S. (1978a) Prolongación de la vida de almacenamiento del nopal hortaliza (*Opuntia inermis* Coulter) por refrigeración. Chapingo, *Neuva Epoca*, **10**, 30–2.

Ramayo, R.L., Saucedo V., C. and Laksminarayana, S. (1978b) Causas de altas perdidas en nopal hortaliza (*Opuntia inermis* Coulter) almacenado por refrigeración y su control. Chapingo, *Nueva Epoca*, **10**, 33–5.

Russell, E.C. and Felker, P. (1985) The prickly pear (*Opuntia* spp): plants with economic potential, in *Management and Utilization of Arid Land Plants*, Symposium Proceedings, Saltillo, Coah., Mexico, pp. 41–6.

Russell, E.C., and Felker, P. (1987a). The prickly pears (*Opuntia* spp. Cactaceae): A source of human and animal food in semiarid regions. *Econ. Bot.*, **41**, 433–45.

Russell, E.C., and Felker, P. (1987b). Comparative cold-hardiness of *Opuntia* spp. and cvs. grown for fruit, vegetable and fodder production. *J. Hort. Sci.*, **62**, 545–50.

Sanchez-Mejorada, E.R.H. (1982) *Algunos Usos Prehispanicos de las Cactaceas entre los Indigenas de Mexico*, SDA, Gobierno del Estado de Mexico.

Sanchez, S.A., Jimenez G., E. and Zarate E.M.S. (1990) Desarrollo de una mermelada a base de nopal adicionada de fruta, in *El Nopal su Conocimiento y Aprovechamiento*, 3a. Reunion Nacional la. Internacional. (eds J.J. Lopez G. and M.J. Ayala O.), Universidad Autonoma Agraria 'Antonio Narro', Buenavista, Saltillo, Coah., Mexico, pp. 291–9.

Sodi, P.E. (1968) Las cactaceas en las epocas precolombinas y virreinal en el siglo XIX. *Cact. y Suc. Mex.*, **XIII** (1), 3–12.

Sutton, G.B., Ting, I.P. and Sutton, R. (1981) Carbohydrate metabolism of cactus in a desert environment. *Plant Physiol.*, **68**, 784–7.

Trejo Gonzalez, A., Calva, E. Monterrubio, P.J. *et al.* (1991) Anti-diabetic properties of prickly pear cactus. Proc. Second Annual Texas Prickly Pear Council, McAllen, Texas.

Uphof, J.C. Th. (1916) Cold-resistance in spineless cacti. University of Arizona Agricultural Experiment Station, Bulletin 7a,119–44.

Williamson, J. (1975) Useful Plants of Malawi, University of Malawi/Montfort Press, Limbe.

7

Hearts of palm (*Bactris, Euterpe* and others)

P.C. Tabora, Jr., M.J. Balick, M.L.A. Bovi and M.P. Guerra

7.1 INTRODUCTION

The heart of palm is the large meristem of diverse palm species. It is pickled and bottled or canned for international commerce, but is often used as a fresh vegetable in areas where the palms are grown. It ranks among the favourite salads of South and Central America and also Southeast Asia. Heart of palm use goes back to pre-Colombian times in the New World. The indigenous tribes had multiple uses for it, i.e., direct consumption of the fresh 'cabbage' or heart itself or the fermentation of the palm cabbage sap as an alcoholic beverage. In Southeast Asia palm cabbages may be pounded and soaked to collect a starchy product that settles to the bottom upon standing. The starch is dried and used for breadmaking. Not all hearts of palm are edible: some are very bitter and some may actually be toxic, for example, the heart of *Orania* sp. (NAS, 1975) used for arrow poisons (Jones, 1984).

In this account, the focus is on the use of the heart of palm as a vegetable. Three major palms are extensively discussed because of their present economic importance: those of the genus *Euterpe* (juçara and açai) from Brazil, those of *Bactris* (pejibaye) of South and Central America and *Cocos* (coconut) in Southeast Asia. All three groups have provided the international market with as much as 25 000 tons of heart of palm valued at about $65 millions as of 1990. Brazil dominates the world market by supplying 85% of the heart of palm – extracted mostly from wild stands in the Amazon and in the southern states of San Paolo, Santa Catarina and Parana but cultivation has begun. Costa Rica has been cultivating pejibaye palms exclusively for heart of palm production and has now some 2000 ha planted. Southeast Asia (mainly Philippines and Thailand) have sporadic

Pulses and Vegetables. Edited by J. T. Williams. Published by Chapman & Hall, 2–6 Boundary Row, London SE1 8HN. ISBN 0 412 46610 4.

plantings of coconuts specifically for heart of palms but most hearts are from old palms being felled for replanting programmes. Large consumers of heart of palm have been France and the USA.

The above three palms typify three diverse types of taste in heart of palm. *Euterpe edulis* (juçara) has a distinct 'bite' favoured by the French market. The juçara has been replaced by *Euterpe oleracea* (açai) which also has a bite although not as strong. *Bactris* has a more neutral taste, although some varieties have a faint bite, a little less than *Euterpe oleracea*. Coconut typifies a sweet type favoured in mixed vegetable sautées and curries.

There are many local names for heart of palm, but the following are the major ones: Florida and Trinidad, palm cabbage or palmetto; South and Central America, palmito; Philippines, ubod; France, coeur de palmier; Spain, corazón de palma or col de palma; Portugal, curaçao do palma; Italy, cafaglione.

7.1.1 **Morphology and anatomy**

All palm hearts can be divided into three parts: the base, the cylinder and the free top (Bernhardt, 1987, Mora-Urpi, 1989a), all of which are edible. The **Cylinder** is a tube-like leafsheath enclosing the less developed petioles and leaflets, so called because it is shaped as an even cylinder until the point where it splits or opens and reveals the rachis and the free leaflets. The **free top** is the composite of the tender rachis and leaflets not enclosed or clasped by the cylinder. The **heart base** includes the enlarged bulbous mass of the tender portion up to the final curve at the beginning of the cylinder immediately after the last exposed heart base.

Palm hearts can be obtained from large mature trees, but the present palm heart technologies emphasize the use of shoots of juvenile plants from recent seeding or of suckers aged 12–36 months depending on the type of palm (Figure 7.1). The recommended size of the shoot to cut is about 7–11 cm whole stem diameter in order to have a standard 1.7–3.5 cm diameter range of the cylinder needed for canning. Single stem palms include *Euterpe edulis*, and *Cocos nucifera*. The multiple stem palms include *Euterpe oleracea*, and *Bactris* sp.

There are many other palms with very little or none of the desired long or evenly round cylinder such as the African oil palm, *Elaeis guineensis*; American oil palm, *Elaeis melanococca*; *Orbigyna* sp.; *Jessenia* sp.; *Acrocomia* sp.; and *Corypha* sp., but they are processed as heart of palm by slicing the bases into appropriate sizes. The African oil palm is notably triangular in shape (Zantua *et al.*, 1989).

The bases of the palm hearts have a coarse appearance with either a vertical or horizontal cut as the unorganized vascular bundles are cut. These resemble the cut section of bamboo shoots. Pejibaye, with its naturally yellowish tint and smaller diameter at the base, has a particular

Figure 7.1

bamboo resemblance. The slight astringent taste and firm texture of the basal pejibaye make it a very good bamboo shoot substitute.

From the tops the rachis is sometimes used as a filler in canning because of its roundness. The leaflets often become disorganized and are set aside for animal feed, but are sometimes pickled.

7.1.2 **Biochemical composition, nutritional and culinary values**

Hearts of palm are relatively rich in protein (Table 7.1) as they are immature meristems, and contain 17 amino acids (Ferreira and Paschoalina, 1987). They are low in fats and sugars and are excellent diet fibre sources (Table 7.2). They are also a moderate source of calcium (Mora-Urpi *et al.*, 1991; Ferreira and Paschoalino, 1987) (Table 7.3).

Pejibaye shows a very low polyphenoloxidase enzyme activity (1.37 units/ml in cathecol of 0.2 M and 4.5 units/ml of guaiacol 0.0072 M) which is advantageous over juçara. Juçara shows a high polyphenoloxidase enzyme activity (41.7 units/ml) and also high peroxidase activity (13.4 units).

Table 7.1 Amino acids for hearts of palm of pejibaye and juçara expressed in g/100 g of dry matter

	Pejibaye	*Juçara*
Lysine	1.09	1.71
Histidine	0.36	0.52
Arginine	1.13	1.76
Aspartic acid	2.18	1.99
Threonine	0.78	0.90
Serine	0.92	1.24
Glutamic acid	1.50	1.56
Proline	0.86	0.99
Glycine	0.80	1.13
Alanine	1.03	1.19
Cystine	1.48	1.89
Valine	0.83	1.22
Methionine	0.28	0.36
Isoleucine	0.84	0.93
Leucine	1.46	1.56
Tyrosine	0.37	0.70
Phenylalanine	0.49	0.70
Tryptophan	0.47	0.35
Essential amino acids	6.24	7.73
Total amino acids	16.87	20.70

Source: Ferreira and Paschoalino, 1987.

Table 7.2 Chemical composition of the hearts of pejibaye and juçara palms

Components	*Pejibaye*[a]	*Pejibaye*[b]	*Juçara*[b]
Water(%)	90.47	88.40	90.81
Crude Protein (%)	2.27	2.32	2.18
Fat (%)	0.13	2.16	2.51
Ash (%)	0.93	1.21	1.35
Crude Fibre (%)	0.89	1.05	0.99
Total sugars (%)	2.7	35.80[c]	12.28
Ascorbic acid (mg/100 g)	14.0	3.20	11.6
Tannin	–	0.80	1.68
Organic acids	–	Lactic acid	Maleic acid
Cyanohydric Acid	–	3.73	0.95
Initial pH	6.58	–	–

Sources: [a] De la Asunción, cited by Mora-Urpi *et al.*, 1991;
[b] Ferreira and Paschoalino, 1987
[c] Expressed on the basis of dry matter of lyophilized material

Table 7.3 Mineral composition of hearts of pejibaye and juçara expressed in mg per 100 g of dry matter

Mineral	*Pejibaye*	*Juçara*
Phosphorus	94.0	68.0
Iron	4.3	3.5
Calcium	114.0	110.0
Magnesium	80.0	63.0
Sodium	1.33	4.17
Potassium	337.6	438.9
Copper	0.159	0.086
Zinc	0.79	1.38
Manganese	0.48	0.61

Source: Ferreira and Paschoalino, 1987.

This difference accounts for the rapid browning of juçara when cut and exposed to air. Pejibaye does not oxidize.

Comparing juçara and pejibaye, there are less reducing sugars in juçara which attests to its non-sweet flavour. This is also compounded by higher tannin in juçara (twice that of pejibaye) which gives the distinct astringent flavour to juçara. To add to flavour distinction, juçara has more maleic acid and pejibaye has more lactic acid (Ferreira and Paschoalino, 1987).

The colour reactions after heat treatment differ. Pejibaye turns creamy or yellowish when cooked and even its carrier solution turns yellowish. Juçara and coconut are whiter when cooked (Ferreira and Paschoalino, 1987). With months of storage pejibaye has a pronounced yellowish tint and juçara and coconut have a pinkish tint.

As to texture, pejibaye is firmer than juçara (Ferreira and Paschoalino, 1987), but this is easily corrected by more cooking time to make the pejibaye softer. The açai is also firmer than juçara. The pejibaye's firmness makes it more crunchy in salads than juçara. Juçara's tenderness or mushiness however, is preferred in some culinary preparations.

The coconut meristem has also been analysed and is also a good protein source (Bondad, 1978) and compares well with cabbage (Table 7.4) not only in protein, but also in carbohydrate. Crude protein can be as high as 11%. The coconut heart oxidizes fast and has to be soaked in a brine solution upon cutting, just like açai and juçara, to arrest oxidation. Its texture is also soft just like the juçara.

7.1.3 **Post-harvest technology and processing**

The heart of palm for international commerce requires few standards, and standards are not very rigidly imposed due to the problems of extraction in Brazil. Standards are, however, the basis for post-harvest practices and

Table 7.4 Nutritional value of coconut compared to cabbage

Value	*Coconut*	*Cabbage*
Protein (g)	4.3	1.10
Carbohydrates (g)	5.6	4.30
Fat (g)	0.8	0.20
Vitamin A (units)	0.0	100.00
Thiamine (mg)	0.5	0.04
Riboflavin (mg)	0.3	0.04
Niacin (mg)	16.0	35.00
Calcium (mg)	33.0	45.00
Iron (mg)	0.2	0.30

Source: Martin, 1984.

technologies. In Costa Rica the requirements for the palm hearts are: (a) white colour, (b) tender with low fibre and (c) a can of 220–225 g drained weight must have at least six sticks (trozos) (Mora-Urpi *et al.*, 1991; Astorga, 1989) of 10 cm long (9 cm long in Brazil).

These qualifications are often re-interpreted and white can mean a range from white to yellow, or does not oxidize to give brown or pink colour. Tender with low fibre can mean that the tissues crumble upon chewing, as opposed to hard which means that the fibres have become too lignified and resist chewing. The requirement of six sticks can mean diameters of 1.5–3 cm as a common range. More slender sticks akin to asparagus spears are now preferred. The ideal shoot or stem size could mean a circumference of 7–10 cm diameter or 20–36 cm circumference so as to give palm hearts of about 2 cm diameter, 40 cm long and weighing 150 g or more.

Upon harvest in Costa Rica the drying external sheaths (cascara I) are stripped from the shoot and left in the field, but one or two protecting sheaths (cascara II) covering the internal hearts are kept to minimize bruising damage during transit to the processing plant. These two protective sheaths are removed in the factory, prior to processing. All shoots must be at the processing plant within 24 h after harvesting. In Brazil as many as four sheaths are left to protect the palm heart due to longer distances of transportation. The quicker the shoots are processed, the higher the yields and the better the quality and the less the microbial contamination. As shoots are stored longer, they become more fibrous. Harvested shoots must also be kept in cool places to minimize dehydration.

Processing

Heart of palm processing is basically a pasteurization treatment to deactivate micro organisms. The heart of palms are sterile at cutting, but the

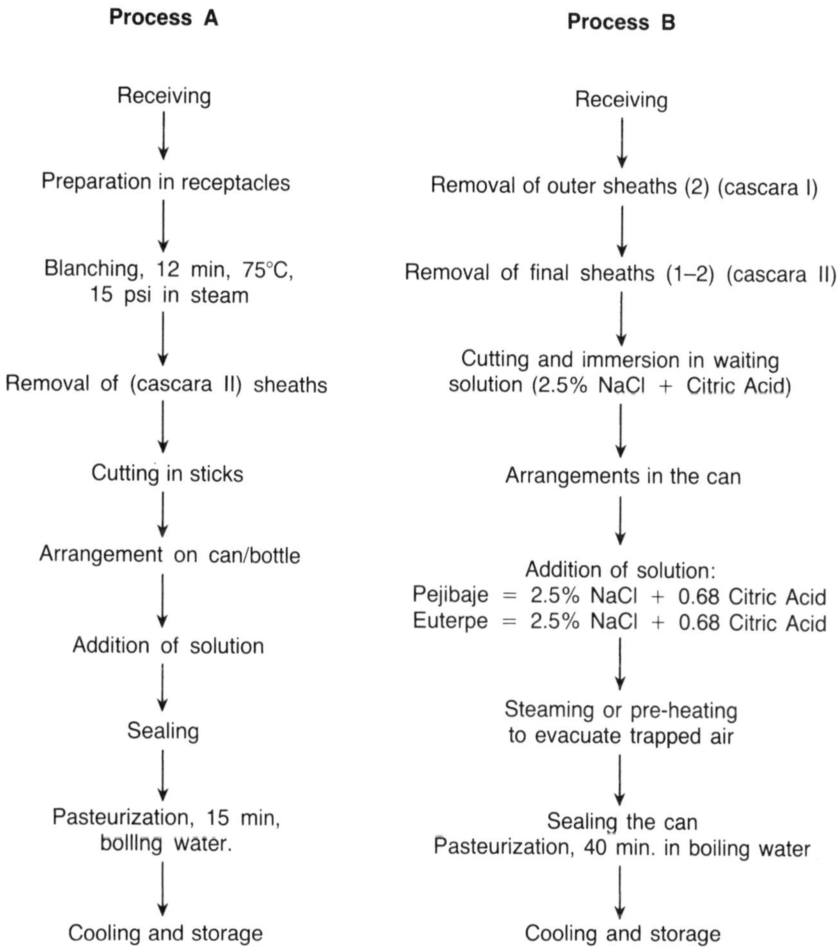

Figure 7.2 Two process flow charts for heart of palm treatment (modified from Ferreira and Paschoalino, 1987).

transport, storage and handling expose them to bacteria and fungi. As the heart of palm for international commerce is canned, pasteurization is carried out to ensure against botulism, a common canning problem (see Figure 7.2).

In Amazonia, small processing plants may have boilers which are fuelwood fired and while water is being heated the cleaners and cutters are working. Process A in Figure 7.2 pretreats the hearts at the beginning whereas process B has two heatings at the end of the process.

For bottling, the pre-heating can be missed from the process. The time of heat treatment can vary depending on the portion of the cylinder. In Brazil the cut portions are grouped into the more tender immature and the

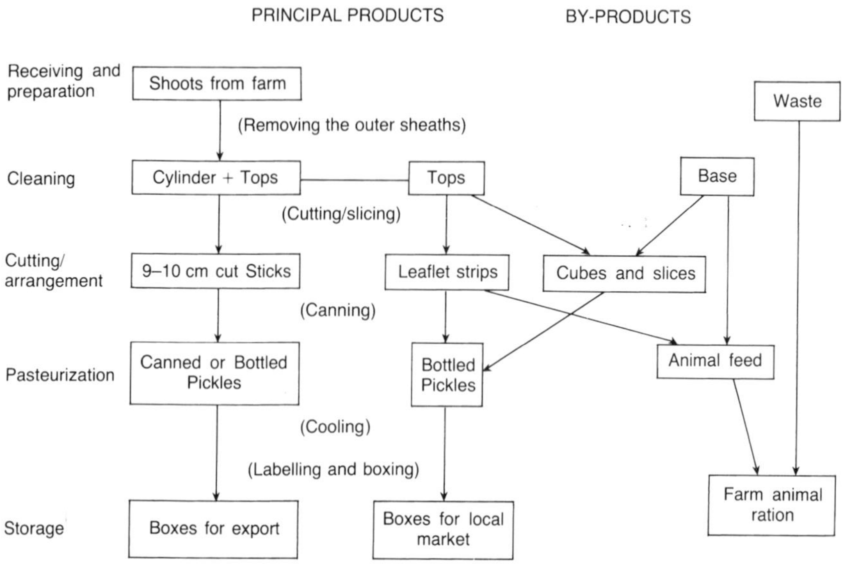

Figure 7.3 Principal products and by-products in processing.

slightly harder and more fibrous parts. The more mature and harder portions receive as much as 40 min in boiling water to soften the fibres. The products and by-products are shown in Figure 7.3.

In Costa Rica the final product is a 24-can box of 12 kg gross weight with a 5.28 kg drained weight of heart of palm. Some bottled hearts are also exported. In Brazil the product could be the same 24-can box of 12 kg gross weight or a 12-can box also of 12 kg heart of palm drained weight. Bottled products of all sorts are sold in the local markets.

Post-harvest and fresh vegetable presentation

Non-pickled heart of palm is widely used in Costa Rica, Southeast Asia, Sri Lanka and India. Thus fresh heart of palm can be obtained in the markets at any time. In C. America, pejibaye cylinders as well as small palm heart bases are packed in plastic bags and refrigerated. Pejibaye can last for 7–8 days in this manner, without any other treatment. In evacuated bags the pejibaye can stay white and fresh for 21 days at 10°C (Tabora *et al.*, 1991). In both cases the fibres progressively develop and palm hearts may become tough.

In Southeast Asia, the coconut is available in the markets as a whole shoot, some very large from mature trees and some very slender from juvenile plants. Restaurants particularly prefer the large ones to use the curved petioles for decoration. In the Philippines the coconut heart is cut into pieces and soaked in brine water to prevent oxidation.

7.1.4 **Trade**

World trade in 1990 reached $65 million. This included Brazil's exports of about $50 million followed by Costa Rica's $6 million (Mora-Urpi *et al.*, 1991). The balance comes from Colombia, Venezuela, Ecuador, Peru, Mexico, Argentina, Thailand and the Philippines. Some 25 000 t are exported from these countries (CECON-OEA, 1977). Trade has been growing erratically due to commercial extinction of jucara in 1985 and 1986 and environmental concerns for the Amazon acai in 1989. Some 85% of all hearts of palm traded come from extraction of primary forest stands. Nevertheless trade has increased in a general pattern since 1972 with shifts into new extraction areas and substitute palm species in Brazil, a gradual increase in plantations in both Costa Rica and Brazil based on pejibaye and an increase in demand. France has been the biggest importer (4500 tonnes), consuming roughly 63% of Brazil's traded heart of palms in 1988. The USA (1000 t), Spain (576 t) and Italy (227 t) are the next big three in growing importance as volumes and value of total imports have tended to increase by about 75–90% annually since 1975. Southeast Asia has decreased its share of the market even though it offers much lower prices. Local markets supply sliced hearts of palm. Many non-exporting countries such as Guatemala, Belize and even the major exporters such as Costa Rica and the Philippines serve basal meristem slices in a myriad of preparations such as sticks, cubes, gratings and decorative slices in pickles, sautées, mayonnaise and vinegar-dressed salads.

Local consumption is high in Brazil, about 80% of its total production or about 80 000 t. All of this is in pickled form, in the cylindrical shape. Local markets in Costa Rica carry the bottled pickled cylinders and also the slices in both bottles and plastic bags. Fresh hearts of palm are neatly packed in transparent plastic bags and show up in supermarket refrigerators. The Southeast Asian markets carry whole fresh shoots, which are used as garnish for local curry and vegetable sautées to substitute for bamboo shoots.

7.1.5 **Farmers' perception of the crop**

Pejibaye farming for palm heart has been increasing, through small and medium-scale farmers, in Costa Rica at a faster rate than with large farms of over 100 ha (Zamora, 1990). The crop's labour intensiveness lends itself to smaller farming schemes. In Colombia (Lozano, 1990; Velasco, 1989), however, large-scale farming is being implemented, using banana farming technologies to transport the stems via cableways to central processing centres.

Brazil, Ecuador and Bolivia have also been increasing their plantations. These are recent plantings (since 1985). Local income compared with coffee and banana puts it in a very attractive position (Tabora, 1987). Its cultural requirements are much less demanding than either coffee or bananas with less problems of pests and diseases as well as less fertilizer needs. The present gross income of $9000/ha/yr is higher than coffee and the market has had less fluctuations. Compared to banana, pejibaye heart of palm show less investment requirement and just as good rates of returns on investment.

There are some problems. Pejibaye needs good drainage and Costa Rican farmers already feel this is a limitation on their farms. It is also estimated that some 76% of the farms have problems with rodents attacking the young shoots.

Juçara and açai farming in Brazil has been recent. The State of São Paulo has already initiated formal plantations. Açai has been growing very well in the São Paulo State (Salles de Aguiar, 1987) and plantations are being developed even while waiting for new hybrids. In the Amazon, açai plantings have been in association with cacao and other crops (Calzavarra, 1987) with açai as a shade tree. Açai has been observed to yield three times more than juçara (Bovi *et al.*, 1987) but this yield has also been superseded by pejibaye (Clement, 1988; Moreira Gomez and Arkcoll, 1987). It seems that heart of palm is quite lucrative in Brazil and both extraction and formal farming are ongoing.

Southeast Asia's coconuts are abundant at all stages. While farming of the juvenile palms for hearts is being done, there is competition of hearts from old plantations under replanting. Thus, farming for palm hearts has not prospered in Southeast Asia except in the vicinities of large urban areas, such as Manila and Bangkok, which provide big markets for the fresh heart of palm.

7.1.6 **Future potential and the role of various palms in different regions**

Heart of palm farming will be a regular feature in the future as palms become scarcer in such places as Brazilian, Venezuelan, Colombian and Peruvian Amazonia where some 85% of all hearts of palm are now being extracted. Farming will also follow the increasing world demand as the heart of palm becomes more accessible. Furthermore as technology is developed palm heart farming could be more productive and lucrative even as prices are reduced. New palms may be brought in for this purpose (Appendix). The royal palm has that potential and recent tests in Honduras show it could yield as much as 5.2 t/ha as a 13 month crop. Granting a 4 ton yield/crop/yr this is more than twice the normal commercial yields of pejibaye. Nevertheless, recent Costa Rican experimental results for pejibaye have also yielded as much as 3 t/yr, twice the present farm yields, and this could already be a competitive crop yield at lower costs (Vargas, 1991). It is expected that pejibaye, juçara, açai and even royal palm will have new technologies as a result of new breeding and cultural managements and they will exceed present yields and profitabilities. Each region will perhaps have a more preferred palm due to edaphic and climatic limitations as well as type of agriculture. In Brazil, an agroforestry scheme combined with the management of existing stands, will become a major need and the model in Santa Catarina may be adapted for this purpose.

It is clear that heart of palm competes with asparagus, bamboo shoots, and certain tubers. While this may be true to a certain extent, the heart of palm has its own distinct appeal. Its slightly astringent taste and crispiness stand out as unique for appetizers. Neither asparagus nor bamboo shoots match it. Asparagus has a chlorophyllous flavour which is not found in the heart of palm. In mixed vegetable salads, the heart of palm will also be useful, but there is no direct substitution function here. Mixing is a matter of form, colour, taste and texture.

New products are expected to increase consumption of heart of palm in the world market, such as slices and juliennes mixed with other vegetables. Fresh rather than canned palm hearts are also expected to enter the world market. Markets in Florida, USA, offer fresh palm hearts and this market can be tapped readily. Florida uses both the royal palm and the sabal palm for this.

The main determinant will be producing the hearts of palm competitively. The Pará, Brazil, industry produces a box of 12kg gross (or drained weight of 6 kg heart of palm) at a cost of \$4.20–4.80 for the raw heart of palm, \$4.20 for canning (labour, materials and energy), \$5.40–6.00 for management overheads and marketing, \$3.60 for amortization and about \$1.00 for taxes. This brings a total cost of \$18.40–19.60 per box. The São Paulo, Brazil, industry pays \$6.20–8.00 per box for the raw material and Costa Rica pays as much as \$9.33 for the equivalent heart of palm (based on \$0.26 per shoot × 1.5 shoots/can of 220 g × 5.28 kg net weight of palm heart/box). Costa Rica trades off with a more efficient and lower cost management due to full integration of production, processing and marketing. No doubt, the future will depend on technological efficiency for Costa Rica and for greater integration for Parã and São Paulo to compete in the world markets. There is also room for improvement in canning and packaging costs. If these happen, hearts of palm can be within reach of a much greater number of people.

7.2 PEACH PALM (PEJIBAYE)

With recent technology development, peach palm has become the pre-eminent heart of palm under cultivation (Zamora, 1989) and has shown agribusiness profitabilities better than those of *Euterpe* sp. trials in Parã

(Moreira and Arkcoll, 1987). Its flavour has been adjusted to customer tastes and as a cash crop it compares favourably with coffee and banana (Tabora, 1987).

7.2.1 **Taxonomy, ecology, and history.**

About eight different species of *Bactris* are recognized at present (Clement, 1988), but full inventory and classification still remains to be done. The pejibayes of Costa Rica are classified as a sub-genus *Guilielma* ($2n = 28$), with four species. It is theorized that the diverse taxa in as many as 11 countries are derived from the Peruvian–Bolivian Amazon's *Guilielma microcarpa*. Clement proposes the following classification:

B. gasipaes Humboldt, Bonpland and Kunth
  Syn. *G. speciosa* Mart.
  Syn. *G. utilis* Oersted
  Syn. *G. chontadoro* Driana
  Syn. *G. microcarpa* Huber = *B. dahlgreniana* Glassman
B. insignis (Mart.) Baillon
B. ciliata (Ruiz & Pavon) Mart.
B. macana (Mart.) Pitter
B. caribaea Kaisten

Peach palm is found in Bolivia (tembé), Peru (pijuayo), Colombia and Ecuador (cachipay and chontaduro), Brazil (pupunha), Panama (pibá), Venezuela (pijiguao), Trinidad (peach palm), Costa Rica and Nicaragua (pejibaye), French Guiana (Paripon) (Pinheiro and Balick, 1987) and Honduras (chúpa o súpa). Most stands occur below 900 m elevation in very humid tropical regions with temperatures above 22°C and rainfall between 2000 and 5000 mm. The palms grow on varied soil types but always with good drainage. They can tolerate acid soils and have high mycorrhiza associations (Clement, 1988).

Peach palm is South America's most utilized native palm and its use is ancient as a source of food (oil and starch) from the fruits and later as a vegetable from the heart of palm. Its multiple use is probably the reason for its widespread distribution and diverse ecotypes (Balick, 1984). Its popularity diminished during colonial times but has recently surged since the 1970s as research proved it to be a good food and the basis for profitable enterprises. New uses have appeared as flour, feedmeal, cooking oil and the trunks have craft value.

7.2.2 **Morphology, growth and development**

The plant is caespitose (clumping) with a fibrous root system and spines cover some internodes but these alternate with nodes without spines.

Table 7.5 Nutrient-removal of heart of palm from pejibaye based on 3200 hills/ha and a production of 9000 heart of palm stems/ha/yr in La Rita, Costa Rica, 1989

	Nutrients removed in kg/ha/yr				*Proportion of nutrients removed*	
Elements	*leaves*[a]	*Gross wt.*[b]	*Total*	*Net wt.*	*Gross wt. (%)*	*Net wt. (%)*
N	503.000	28.000	531.000	8.650	5.270	1.630
P	33.100	4.800	37.900	1.340	12.660	3.530
K	217.300	31.000	248.300	5.760	12.480	2.320
Ca	60.100	4.700	64.800	0.880	7.250	1.360
Mg	39.100	3.900	43.000	0.770	9.070	1.790
Fe	1.200	0.030	1.830	0.006	1.640	0.330
Cu	0.163	0.021	0.184	0.004	11.410	2.170
Zn	0.204	0.050	0.254	0.011	19.680	4.330
Mn	2.189	0.085	2.274	0.014	3.740	0.620
S	43.870	3.360	47.230	0.672	7.110	1.420
B	0.528	0.029	0.557	0.004	5.210	0.720
Cl						
Al						

[a] leaves and outer sheaths
[b] net weight of heart of palm + by-products + inner sheaths.
Stem diameters at 8–9 cm.
Source: Herrera, 1989.

Leaves are large, 2 m in length and with 200 leaflets or more. Every axil gives rise to a shoot or a flower bunch. When cut at the base, shoots are formed. When the tree is left to grow tall, there is no sucker profusion. Thus, when cultivation is for fruits, it cannot be a heart of palm source.

Seedbeds can be used to germinate seeds, but direct planting in bags is also recommended. Germination begins 45 days from seeding and concludes after 60 days. At present tissue culture plantlets are not yet available, thus seeds are used extensively. Sucker separation has been tried (Blaak, 1980) but this is too cumbersome and less successful. Beds and bags must be moist and seeds must have at least 40% humidity. Germination can range from 75%–99% with better results in bags.

Planting distances are at 2 × 1 m or 1.5 × 1.5 m (Mora-Urpi 1989c). At six months plants are ready for transplanting. At 12–18 months after transplanting the first harvest can be done. Subsequently three stems can be harvested per clump/year giving about 15 000 stems harvested/ha/year. Six stems are allowed to grow at any one time.

The first eight months after establishment require rigorous weed control as the field is open. After eight months the leaves begin to close up and weeding becomes minimal. Fertilization depends on soils and climate but as a general guide, heart of palm production requires high nitrogen. The nutrient removal (Table 7.5) and recycling scheme are good bases for fertilizer recommendations.

For the Atlantic coast of Costa Rica where the soil has been analysed (Herrera, 1989; Sancho and Lopez, 1990) the recommendations for fertilizers/ha/yr are 200–220 kg N; 20 kg P_2O_5; 160–200 kg K_2O 50–100 kg Mg; and 400–500 kg Ca.

7.2.3. **Pests and diseases**

The four major pests in Costa Rica are:

1. the sugarcane weevil (*Mitamasius hemipterus*);
2. the basal stem beetle (*Strategeus aloeus*);
3. the foliage mite (*Retracus johnstonii*) (Mora-Urpi *et al.*, 1984);
4. the nibbler (*Orthogeomys cherriei*) (Delgado, 1990).

The best ways to avoid these pests are by ensuring there is no sugarcane or banana nearby (for the weevil), applying a chlorinated insecticide on the holes and closing them (for beetle), and spraying wettable powder of sulphur (3–4 lbs/50 gal water) for the mite.

The diseases of pejibaye are:

1. *Pestalotiopsis* sp., oval or round yellow spots on leaves having a necrotic appearance;
2. *Mycosphaerella* sp., round light brown spots surrounded by dark brown borders and by a well-marked yellow border on young and old leaves;
3. *Colletotrichum* sp., irregular spots surrounded by small yellow borders for young seedlings and these lesions open up the way for *Erwinia* sp., which cause rotting of medula and give a bad odour; and
4. *Phytophthora* sp., rotting of the base of leaves which turn yellow at the tips and also produce a bad smell (Mora-Urpi *et al.*, 1984; Vargas, 1989).
5. A nursery disease of seedlings in Brazil is *Ceratocystis paradoxa* (Bovi *et al.*, 1987).

7.2.4 **Yields and economic profile**

Recent advances have shown that pejibaye can yield 1.76 t/ha (Mora-Urpi *et al.*, 1991) in commercial fields. As younger shoots and smaller hearts are now preferred, cultivation is also changing. With higher densities, experimental yields have gone up to 3 t/ha/yr (Zamora, 1990), beginning the third year. Cannable yields have been only 30%, but it is only a matter of time for better cannable yields with improvement in cultural management. If this happens, pejibaye for heart of palm will be a very lucrative crop.

In 1984 the estimates on gross income for pejibaye heart of palms was \$4400/ha. This had doubled to \$9000/ha in 1992 even when prices dipped

40%. The increase is due to new technologies allowing the harvesting of thrice more stems from the same area. With $9000 per ha/year gross income, heart of palm is already earning more than coffee.

7.2.5 **Genetics and breeding**

Three major germplasm collections exist in Brazil (Manaus), Colombia (Vallecuacano) and Costa Rica (Guapiles) and they should be good bases for breeding. Costa Rica has organized an intensive research programme breeding palms for greater vigour (faster growth) and longer cylinders for hearts of palm. The Putumayo race (Mora-Urpi, 1989b) with less spines is close to the ideal, but its suckering has to be improved. At present there are data on allozyme variation of pejibaye which should be a basis for breeding. The University of Costa Rica and the Organization for Tropical Studies have a collaborative project with the University of Washington for this purpose.

7.3 *EUTERPE EDULIS* (JUÇARA) AND *EUTERPE OLERACEA* (AÇAI)

7.3.1 **Taxonomy, ecology and history**

There are 29 species of the genus *Euterpe* (Bovi *et al.*, 1987). Three are important ones for palm hearts. *E. edulis* Mart. is a single-stemmed palm distributed along the southern seacoast of Brazil (São Paulo, Santa Catarina and Paranà). *E. oleracea* Mart. occurs in the Amazon region of Brazil and *E. precatoria* occurs on the Peruvian-Brazilian-Colombian borders (Calzavarra, 1987 and personal communication).

The two most important Euterpes are juçara and açai. Juçara has been decimated in its natural stands in Brazil due to over gathering for commercial purposes and the extreme popularity of the heart of palm that led to such excessive extraction (Arkcoll and Clement, 1989). Juçara plantations have been slow in coming as juçara is a slow grower. Its need for shading means that it cannot be grown alone in plantings but needs a form of agroforestry not yet developed. In natural vegetation in 1987, *E. edulis* had a density of 0.66 palms per m^2 in different stages of development (Bovi *et al.*, 1987) with some 620 harvestable shoots/ha (Kirchner *et al.*, 1987). The palm can live up to 50 years (Bovi *et al.*, 1987) and produce 3000 seeds from three inflorescences per year beginning in the sixth to eighth year. The most critical period for survival seems to be to the first 50 cm stage when young plants compete for sun, water and nutrients.

The climate of the São Paulo seacoast, where there are large juçara populations, has high rainfall (1400–2800 mm), with temperature ranging from 20–23°C. On the more elevated (700–900 m) ranges of São Paulo State, the rainfall is 1300–1500 mm, but is well distributed, with temperature about 18–20°C. In places with high humidity some populations exhibit 6–11 palms per m^2 density. The soils where these palms are located are low in phosphorus, K, Ca and Mg with pH 4.1–5.6 and high organic matter.

7.3.2 **Morphology, growth and development**

Juçara germination takes up to six months to complete. The seeds have a hard endocarp that inhibits the penetration of water to the embryos. Depulping and soaking the seeds for two days reduce the germination time to 35 days and increase the percentage of germination from 53.19 to 91% (Bovi *et al.*, 1987). Açai is faster to germinate (30 days) and loses viability fast (Calzavarra, 1987). Seed storage is possible for juçara up to five months by bringing down the temperature to 5–10°C and raising humidity. Best results are obtained by using hermetically sealed containers with seeds fully imbibed at 5–10°C (Bovi *et al.*, 1987).

Seeding is done directly in plastic bags or in shaded beds for germination, and transplanting is done before or after one leaf has developed (Nodari *et al.*, 1987). Direct planting is not recommended (Calzavarra, 1987). The germination beds are made of four parts organic matter, three parts compost, one part fine sawdust and one part ash, mixed together.

Recent techniques using somatic embryogenesis in tissue culture have shown that it is possible to mass propagate clones of *E. edulis* (Guerra and Durzan, 1992), but there is still a long way to go before the technology is down-streamed (Guerra *et al.*, 1987). General protocols for the establishment of *E. edulis* were proposed by Guerra and colleagues in 1987.

Phosphorus and potassium fertilization increase the number of leaves, girth and height of palms. Foliar nitrogen fertilization is not recommended as it accentuates leaf enlargement which makes plants susceptible to water stress. Apparently, mycorrhizas are involved in juçara seedlings and they are of the vesicular-arbuscular type; although this is not readily observed in açai (Bovi *et al.*, 1987), they are important (Balick, unpublished).

Larger plants 50–100 cm in height survive better than 10–20 cm (two-leaved) plants during transplanting. Bare root planting of large seedlings does not do as well as roots with soil. Plants with pruned leaves survive better than those with full leaves. Shaded plants have a higher survival (99%) than those with half shade (89%). Planting distance is recommended at 1.5 × 1 m and 1 × 1 m to obtain yields of 2.905 tons and 2.496 t/ha respectively (Bovi *et al.*, 1987) after three years.

Temporary shading is provided through temporary intercropping; permanent shading means an associated crop or a mixed culture system. Bananas are a good temporary shade for juçara for three years. *Gmelina arborea* and *Mimosa scarabella* are good permanent shades. Interplanting with *Hevea brasiliensis* also seems viable (Bovi *et al.*, 1987). For *Euterpe oleracea* shading does not seem to be a major requirement.

7.3.3. **Pests and diseases**

The most important insect pest is *Rhyncophorus* sp. which deposits its eggs on the young unopened leaves. The larvae eat the tender leaves and proceed toward the meristematic apex, killing the plant (Bovi *et al.*, 1987). A fungus disease, *Colletotrichum* sp., attacks açai seedlings and losses can be up to 70% in nurseries. Hybrids have demonstrated resistance to the disease (see below).

7.3.4. **Breeding**

At present a breeding programme at Campinas, São Paulo, has achieved a hybrid between *Euterpe oleracea* and *E. edulis*. (Bovi *et al.*, 1987). This is now being evaluated and initial results point to improved hybrids, with less susceptibility to anthracnose (*Colletotrichum gloeosporioides*) especially in the seedling stages; however, anthracnose is easy to control with fungicides such as Bentlate and Dithane.

7.3.5 **Yields**

The yield from plantings of juçara, açai and the hybrid, in São Paulo three years after planting, are shown in Table 7.6.

A yield of 2.95 t/ha at three years for juçara means 1 t/ha/year of commercial heart of palm. The hybrid (also three years old) has twice the yield with 6.15 t/ha or 2.05 t/ha/year. Açai yield is not easy to calculate. It suckers and needs more time to allow the suckers to be harvested over a long period.

7.3.6 **Exploitation of wild stands and agroforestry**

With the rapid decimation of juçara in Southern Brazil, the estuary of the Amazon has given rise to the new palm heart industry based on açai (Calzavarra, 1987; Strudwick and Sobel, 1988). Açai has long been used to

Table 7.6 Yield indicators of 3-year-old *Euterpe* at Campinas, San Paolo

Indicators	*Juçara*	*Açai*	*Hybrid*
Gross weight of stem (kg)	9.07	4.14	12.59
Net weight of cylinder (g)	443.00	132.43	923.29
Weight of base (g)	146.57	100.85	437.57
Length of cylinder (cm)	41.42	33.57	51.29
Diameter at middle (cm)	3.51	2.25	5.00
Number of cylinders/ 0.5 kg can	1.01	3.04	0.48

Adapted from Bovi *et al.*, 1987.

produce a popular beverage and the Indians of the estuary collect the fruits from protected palms around their houses and concessions. Thus, fruiting trees are not felled. Açai has suckers which are collected when 3–5 years old and big enough to produce 100–150 g cannable cylinders. The buyers normally tell the local people that they wish to buy the açai palm shoots and will pick them up for delivery, usually the following morning, to transport to the processor. (When the tide is low the delivery is postponed until the river swells.) As the shoots can be already three days old, portions tend to become fibrous and this has led to complaints of 'hard' palm hearts. This is corrected by segregating the harder portions for more heat treatment to soften the fibres.

This exploitation of açai by the Amazon natives for dual use has become more stable (Anderson and Jardin, 1989) but the increase in demand and number of processors poses a threat to the fragile ecological balance. A sudden rise in prices of the palm heart shoots could mean the felling of older trees meant for fruit production and general scarcity for a number of years. Thus there is a need to develop schemes for sustainable exploitation and renewal of the populations using agroforestry schemes. There are some trials being done near Florianópolis in Brazil to regulate harvesting and to incorporate some replanting under forest conditions. The scheme is now over 16 years old and seems to be working, but the model is built on a large (5000 ha) hacienda.

Schemes for smallholders need to be developed. Anderson (1988) studied the commercial products sold in a year by a simple family living in the Ilha das Onças region near Belem, Brazil. He found that some 35 026 hearts of *E. oleracea* were extracted during this time, representing $US 2917 in income. At the same time the family harvested $15 533 worth of fruits from this same species (78 885 kg of product). Together these two products accounted for 75% of the product sold by the family. Management systems derived by local people allow for the sustainable production of palm heart in this region of the world.

7.4 COCONUT (*COCOS NUCIFERA*)

Pickled, canned hearts enter international commerce. These are hearts from very old palms (100 years or so) being replaced with new plantings in the Philippines and Thailand (Tabora, 1976). Much of the information on agronomy and yield is available in the literature and is not repeated here.

7.4.1 **Taxonomy, ecology and history**

Cocos nucifera is a single species with many variants and is an ancient cultigen. The most preferred hearts of palm are from the green varieties which are sweeter, whiter and oxidize less. Yellow varieties are less sweet, have an off-white colour and oxidize more. In Southeast Asia the coconut covers large tracts of lands (3 million ha in the Philippines and another 3 million in Indonesia) on diverse soils and varying climates, normally along seacoasts. Some coconuts are even found in swamps, but water-logging makes coconut unproductive for fruits, the reason they were domesticated.

7.4.2 **Growth and Development**

Near large urban areas of Manila and Bangkok, coconuts can be found as small dense plantings of plants very close to each other: 25 × 100 cm or to 50 × 200 cm. The large seednuts are pre-germinated. The ground is simply cleared (no other land preparation) and the nuts lined up one after the other with the shooting edge upwards. It is not necessary to bury the germinated nuts, but, to secure them, they are half-buried by opening a 12 cm trench and lining up the coconuts at spacing of 25 cm.

Experimental plantings at FHIA in Honduras showed that coconut can produce 2.24–3.63 tons of heart of palm (Zantua *et al.*, 1989) in one year taking into account both the cylinder and the basal heart. About 40% of the weight is the industrial cylinder but this does not possess a structure useful in canning.

In La Lima, Honduras coconuts fertilized with 400 kg of N/ha/year respond significantly better than those with 200 kg of N/ha/year. The palm with 600 kg/ha/year did no better than those with 400 kg/ha/yr (Table 7.7).

Table 7.7 Yield (t/ha) of 18-month-old coconuts, 12 months from transplanting with nitrogen fertilization, La Lima, Honduras, 1989, at three densities

	Nitrogen (kg/ha/yr)			*Effect of density*
Plant density/ha	*200*	*400*	*600*	*(average)*
	Aerial parts (t/ha)			
10 000	110.5	103.0	126.0	113.2[a]
20 000	136.5	161.5	161.0	153.0[b]
30 000	187.5	207.0	210.0	201.5[c]
Effect of N	144.8[d]	157.2[e]	165.7[e]	
	Heart of palm yield (t/ha)			
10 000	1.68	1.99	2.44	2.04[a]
20 000	2.35	2.77	2.77	2.63[b]
30 000	3.05	3.53	3.63	3.40[c]
Effect of N	2.36[d]	2.77[e]	2.95[e]	

Randomized block with four replicates. Interaction of density and nitrogen not significant; averages with the same superscripts are not significantly different (0.05).
Source: Tabora and Ramirez, 1991.

7.4.3 **Utilization**

Coconut hearts are used as a fresh vegetable slice with flavoured dips in the Philippines. They are also julienned in Indonesia for the peanut-flavoured gado-gado. In Thailand the petiole base is used to carve designs for sautéed vegetables. In Malaysia slices go into curried vegetables. A favourite dish in Southeast Asia, the eggroll, can be filled with heart of palm julienne mixed with shrimps and other vegetables (called lumpia in the Philippines) and served as 'the millionaire's salad' (Jones, 1984).

APPENDIX: OTHER PALMS USED AS HEART OF PALM

(a) **The royal palms (*Roystonea* sp., previously *Oreodoxa* sp.)**

In Florida, the Caribbean and northern South America, local people have used the royal palm as a source of palm heart (aside from the sabal palm) in traditional preparations (Jones, 1984; Uhl and Dransfield, 1987) for clan gatherings. There are many wild growths of royal palms in Florida.

Experiments in La Lima, Honduras, show the royal palm with high net cannable yields 4.0–5.2 t (cylinder) under high densities of 20 000, 40 000 and 60 000 palms per ha, harvesting the juveniles in 13 months. This is a yield record already and could still be improved, possibly up to 6 t per ha of 10–11 months juveniles. The heart does not oxidize and is very white and the cylinder weighs between 80–250 g, with 2 cm diameters and about 23–42 cm length. Its texture is firm, very much like that of pejibaye, and also has a neutral taste. The apical meristem has a faint astringent flavour like that of pejibaye which, if included in pickling, imparts that particular flavour to even the neutral portions. The hearts are glossy and have an appealing neat presentation.

The architecture of juvenile *Roystonea elata* allows it to be an excellent industrial source of heart of palm. Its disadvantage is that it grows so fast it has to be harvested on time before its diameter grows too big. It is also a non-suckering plant and has to be planted as an annual crop, unlike pejibaye and açai, which are suckering perennials and have smaller diameters. The royal palm's excellent colour, processing properties and precocity make it a very good addition to the palm heart alternatives (Tabora *et al.*, 1991).

Another species of royal palm, *Roystonea oleracea*, from Central America is also a good candidate. It has perhaps greater adaptability in areas with poorer drainage as it normally grows robustly even in water-logged areas in Honduras. It also grows in sandy soils near coastal areas of the Caribbean Islands, Honduras and Costa Rica.

A mature fruit bunch of the royal palm can contain as many as 10 000 seeds which germinate in 30–60 days if seeded immediately. Dried seeds can be stored for two years, but take 6–8 months to germinate. The seedlings have deeper roots than most other palms (Tabora and Ramirez, 1991) and the seedlings grow rapidly.

The most common diseases are those of *Helminthosporium* sp. which attacks the seedlings and *Colletotrichum gloeosporioides* (anthracnose) which attacks the juveniles. Spraying regularly with copper fungicides helps to control the spread of infection. Tilt has been effective to clean off the fungus at this stage. Pests are few, but rabbits are known to attack small seedlings, so protection is necessary.

In fertilizer experiments in Honduras on royal palm, there were no significant differences between 200 kg/ha, 400 kg/ha and 600 kg/ha of N. This is attributed to the extensive root system of royal palm.

As the royal palm stems are smooth, they are easier to harvest, clean and process. On the whole they are also easy to cut. The leaves are upright and there are no suckers that interfere in the operations. In the experiment the yield indicators were as follows:

Plants/ha	*Green mass (t)*	*Weights removed from field (t)*	*Net heart of palm (t)*	*Length (cm)*	*Diameter (cm)*
20 000	138.9	31.7	4.0	31.1	25
40 000	162.0	31.1	4.5	23.8	2.0
60 000	155.0	40.4	5.2	21.7	1.8
cv	11.3	10.3	11.0	8.6	6.3

cv, coefficient of variants.

(b) *Archontophoenix alexandrae* Wendl

This is also a fast-growing palm with a more slender structure than the royal palm. The heart oxidizes slightly, but not much and also has the characteristic astringency very much like the juçara. The dimensions are also just like the juçara, and it is also single stemmed. This palm is much more adapted to drier soils in high rainfall areas, but it is profuse on river banks of Australia and New Zealand.

(c) **The rattans (*Calamus* sp. and *Daemonorops* sp.)**

These are among the most astringent hearts of palms. Nevertheless they are eaten as specialities in the Philippines and Malaysia. The scarcity of the rattans have also made the hearts scarce, but rattans are beginning to be cultivated and the heart may become a by-product. Because of its strong astringent flavour, the rattan hearts may have restricted usage.

Table 7.8 Other palms with edible heart of palm (palm cabbage)

Species	*Origin*	*Taste*
Acanthophoenix rubra	Mascarene Island	Sweet
A. alexandrae	Australia	Neutral
A. cunninghamiana	Australia	–
Acrocomia mexicana	Mexico, C. America	Sweet
A. vinifera	C. America	Sweet
Areca catechu	Asia	Astringent
Arecastrum romanzoffianum	S. America	–
Arenag ambong	Philippines	Sweet
A. listeri	Christmas Island	–
A. pinnata	Indonesia	Sweet
Astrocaryum javari	Brazil	Neutral
A. murumuru	Brazil	Neutral
A. vulgare	Brazil	–
A. standleyanum	Honduras	Sweet
A. tucuma	Brazil	–
Borassodendron machadonis	Malaysia	Sweet
Borassus aethiopium	Tropical W. Africa	–
Brahea brandeegei	Mexico	–
B. serrulata	Texas and Florida	Sweet
Carpentaria acuminata	Australia	–
Chamaedorea elegans	C. America	Astringent
Chamaerops humilis	Italy	Sweet
Corypha elata	Asia	Neutral
C. gebanga	Asia	Sweet
Deckenia nobilis	Seychelles	–
Gronophyllum ramsayi	Australia	–
Gulubia palauensis	Palau Islands, Carolinas	–
Heterospathea elata	Amboina	–
Hyophorbe lagenicaulis	Mascarene Islands	–
H. verschaffeltii	Mascarene Islands	–
Jessenia bataua	Brazil	Neutral
Juania australis	Juan Fernandez	–
Licuala paludosa	Malaysia, Thailand	–
Linospadix monostachya	Australia	–
Livistona australis	Australia	–
L. benthamii	Australia	–
L. humilis	Australia	–
L. rotundifolia	Malaysia, Philippines	Neutral
L. speciosa	India, Myanmar, Malaysia	Neutral
Lodoicea maldivica	Maldive Islands	–
Maximiliana martiana	Brazil	Sweet
Metroxylon sagu	Melanesia, SE Asia	Sweet
Normanbya normanbyi	Australia	–
Oenocarpus bacaba	Brazil	–
Oncosperma horridum	Malaysia, Indonesia	–
O. tigillarium	Malaysia	–
Orbignya cohune	C. America	Neutral
Phoenix canariensis	Canary Island	Sweet
Pritchardiopsis jennencyi	New Caledonia	–
Raphia vinifera	Tropical Africa	Sweet
Ravenea madagascariensis	Madagascar	–
Rhopalostylis sapida	New Zealand	–
Sabal palmetto	USA	Sweet
S. texana	USA	Sweet
Satakentia liukiuensis	Ryukyu Islands	–
Scheelea martiana	Brazil	–
Syagrus siagrus	Brazil	Bitter

(d) **The fish tail palms (*Caryota mitis* Lour, *C. urens* L. and *C. cumingii* Lodd.)**

These have slightly astringent flavours too strong for most people but they are eaten in some regions of India, Malaysia and the Philippines. *C. urens* is a favoured cooking heart of palm. The caryotas are very fast growing and have ideal sizes and length. The cylinders are long and even. Aside from the slight bitterness the caryotas oxidize very rapidly upon exposure to air. Thus they are generally blanched prior to processing. As they are also cooked again during pasteurization, the end result is a mushy heart of palm.

Caryota mitis is a suckering species.

(e) **Sago palm (*Metroxylon sagu* Rottb.)**

This is a large palm which suckers. Often the suckers are pruned and these give hearts of palm with a very sweet taste. The conical architecture of the heart does not lend itself to major industrialization, but its particular flavour may give it a special place among hearts of palms.

(f) **The other palms**

There are probably as many as 60 other palms which can be useful for heart of palm production; some of these are shown in Table 7.8, adapted from Jones, 1984; with additions from Balick, 1982, 1985, 1986, 1987, 1989; Blombery and Rodd, 1982; Dickson, 1978; Dransfield *et al.*, 1988; Johnson, 1986; Martin, 1984; Schery, 1956; Tabora, 1989; Tucker, 1988 and by the authors.

REFERENCES

Anderson, A. (1988) Use and management of native forests dominated by açai palm (*Euterpe oleracea* Mart.) in the Amazon estuary. *Advances in Economic Botany*, **6**, 144–54.

Anderson, A. and Jardin, M.A.B. (1989) Costs and benefits of floodplain forest managements by rural inhabitants in the Amazon Estuary: A case study of the açai palm production, in *Fragile Lands of Latin America*, (ed J. Browder), Westview Press, Boulder, Colorado, pp. 114–29.

Arkcoll, D.B. and Clement, C.R. (1989) Potential new food crops from the Amazon, in *New Crops for Food and Industry*, (eds G.E. Wickens, N. Haq and P. Day), Chapman and Hall, London, pp. 150–65.

Astorga, C. (1989) Producción de palmito de pejibaye en altas densidades y palmeras promisorias para la explotación de palmito, in *Proceedings of the Consultative Meeting on Palm Hearts*, FHIA, La Lima, Honduras, pp. 15–26.

Balick, M.J. (1982) Palmas neotropicales. *Interciencia*, **7** (1), 25–9.

Balick, M.J. (1984) Palms, people and progress. *Horizons*, **3** (4), 32–7.

Balick, M.J. (1985) Useful plants of the Amazonia, in *Amazonia*, (eds G.T. Prance and T.E. Lovejoy), Pergamon Press, Oxford.

Balick, M.J. (1986) Systematics and economic botany of the *Oenocarpus-Jessenia* (Palmae) complex. *Advances in Economic Botany*, **3**, 1–140.

Balick, M.J. (1987) The palm heart as a new commercial crop from Tropical America. *Principes*, **20**, 24–8.
Balick, M.J. (1989) Native neotropical palms. A resource of global interest, in *New Crops for Food and Industry*, (eds G.E. Wickens, N. Haq and P. Day), Chapman and Hall, London, pp. 323–32.
Bernhardt, L.W. (1987) Procesamiento do palmito de javari (*Astrocarium javari*), ler Encuentro de Pesquisadores de Palmitos ANAIS, Curitiba, Brazil.
Blaak, G. (1980) Vegetative propagation of pejibaye (*Bactris gasipaes* H.B.K.). *Turrialba*, **30**, 258–61.
Blombery, A. and Rodd, T. (1982) *An Informative Practical Guide to Palms of the World*, Angus and Robertson Publ., UK.
Bondad, A. (1978) Analysis of coconut ubod (heart), U.P. at Los Banõs, Laguna College, Dept. of Horticulture, mimeographed.
Bovi, M., Godoy, G. Jr. and Saes, L.A. (1987) Pesquisas comos generos *Euterpe* y *Bactris* no Instituto Agronomico de Campinas, ler. Encuentro de Pesquisadores de Palmito ANAIS, pp. Curitiba, Brazil.
Calzavarra, B.B.G. (1987) Pupunheira é Açaizeiro. Recomendacões basicas, Centro de Pesquisa Agropecuária do Tropico Umido, EMBRAPA-UEPAE, Belem, Brazil.
CECON-OEA. (1977) Corazones de palmito, perfiles de mercado, Secretaria General de la Organizacion de los Estados Americanos, Washington DC.
Clement, C. (1988) Domestication of the pejibaye palm (*Bactris gasipaes*): past and present. *Advances in Economic Botany*, **6**, 155–74.
Coyle, L.P., Jr. (1982) *The World Encylopaedia of Food*, Facts on File Inc., New York.
Delgado, R.M. (1990) La taltuza (*Orthogeomys cherrieri*) como plaga de cultivos de pejibaye. *Serie Técnica Pejibaye*, UCR, Boletín Informativo, **I** (1).
Dickson, J. (1978) *Check-list and Uses of Plants in the Wilson Popenoe Botanical Garden*, Ministerio de Agricultura, Honduras.
Dransfield, J., Johnson, D. and Synge, H. (1988) *The Palms of the New World*, IUCN, Gland, Switzerland.
Ferreira, V.L.P. and Paschoalino, J.E. (1987) Pesquisa sobre palmito no Istituto de Tecnologia de Alimentos, in ler Encuentro de Pesquisadores de Palmito, ANAIS, Curitiba, Brazil, pp. 45–62.
Guerra, M. and Durzan, D. (1992) Biotechnological strategies for mass clonal propagation and germplasm conservation of Brazilian pine and palm species, Networking Meeting of the BOSTID, National Research Council of USA, Proceedings of the Workshop, EARTH, Costa Rica.
Guerra, M., Lima da Silva, A. and da Costa, R.M.B.F.L. (1987) Establecimiento de métodos para micropropagacão do palmiteiro *E. edulis*, ler Encuentro de Pesquisadores de Palmito ANAIS, Curitiba, Brazil.
Herrera, W. (1989) Fertilización del pejibaye para palmito. *Serie Técnica Pejibaye*, UCR, Boletín Informativo, **I** (2), 4–8.
Johnson, D.V. (1986) Multipurpose palm germplasm, in *Multipurpose Tree Germplasm*, (eds J. Burley and P. von Carlowitz), ICRAF, Nairobi, pp. 249–78.
Jones, D. (1984) *Palms in Australia*, Reed Books, NSW.
Kirchner, F.F., Lozoya, J.C.R. and Ohlson, J.C. (1987) Quantitativos na estimativa do peso y distribucão por clase do palmito (*E. edulis*, Mart.), ler Encuentro de Pesquisadores de Palmito ANAIS, Curitiba, Brazil.
Lozano, P. (1990) *El Palmito, Corazón de la Palma*, Agricultura de las Americas.
Martin, F. (1984) *Handbook of Tropical Food Crops*, CRC Press, Boca Raton, FL.
Mora-Urpi, J.E. (1989a) Normas de calidad del palmito. *Serie Técnica Pejibaye*, UCR, Boletín Informativo, **I** (1).

Mora-Urpi, J.E. (1989b) Cultivares de pejibaye para palmito. *Serie Técnica Pejibaye,* UCR, Boletín Informativo, **I** (1).
Mora Urpi, J.E. (1989c) Densidades de siembra para la produccion de palmito. *Serie Técnica Pejibaye,* UCR Boletín Informativo, **I** (1).
Mora-Urpi, J.E., Bonilla, A. Clement, C. and Johnson, D. (1991) Mercado internacional de palmito y futuro de la explotación salvaje vs. cultivado, *Serie Técnica Pejibaye,* UCR, Boletín Informativo, **III** (1–2), pp. 6–25.
Mora-Urpi, J. E., Vargas, C.A. López, M., *et al.* (1984) *The Pejibaye Palm (*Bactris gasipaes *H.B.K.),* FAO, Rome, Italy.
Moreira Gomez and Arkcoll, D.B. (1987) Estudios iniciais sobre a produção de palmito de pupunha. 1er Encuentro Nacional de Pesquisadores de Palmito, ANAIS, Curitiba, Brazil.
National Academy of Sciences (1975) *Underexploited Tropical Plants with Promising Economic Value,* NAS, Washington, D.C.
Nodari, R.M., Guerra, P. Reis, A. and Sedrez, M. (1987) Eficiencias e sistemas de implantacão do palmiteiro em mata secundaria, ler Encuentro de Pesquisadores de Palmito, ANAIS, Curitiba, Brazil.
Pesce C. (1985) (as translated by D.V. Johnson) *Oil Palms and other Oilseeds of the Amazons.* Reference Publications, Inc., Algoma, Michigan.
Pinheiro, C.U. and Balick, M.J. (1987) Germinacão de sementes de palmeiras, Depto. de Difusão de Tecnologia, Brasilia.
Salles de Aguiar, C. (1987) Contribucão para a implantacão de cultura do acaizeiro (*E. oleracea*) no litoral paulista, ler Encuentro do Pesquisadores de Palmito, ANAIS, Curitiba, Brazil.
Sancho, H and Lopez, A. (1990) *Observaciones sobre la distribution radical de pejibaye (*Bactris gasipaes *H.B.K.) para palmito en un andosol.* CORBANA, San José, Costa Rica.
Schery, R.W. (1956) *Palmas y Plantas Utiles al Hombre,* Salvat Editores, S.A. Barcelona.
Strudwick, J. and Sobel, G.L. (1988) Uses of *Euterpe oleracea* Mart. in the Amazon estuary, Brazil. *Advances in Economic Botany,* **6**, 225–53.
Tabora, P.C. (1976) Cocobud, the classy vegetable for all seasons. Univ. Philippines, Los Baños, College Laguna (mimeographed).
Tabora, P.C. (1987) *Evaluation of 25 crops for Diversification in Honduras,* FHIA, La Lima, Honduras.
Tabora, P.C. (1989) The utilization of palm hearts in the Asian region, in *Proceedings of the Consultative Meeting on Palm Hearts,* FHIA, La Lima, Honduras.
Tabora, P.C. and Ramirez, T. (1990) Tecnologias de germinacion en el semillero de palma real. Informe Anual Programa de Diversificacion, FHIA, La Lima, Honduras.
Tabora, P.C. and Ramirez, T. (1991) Efectos de poblaciones y niveles de nitrogeno sobre el crecimiento, rendimiento y calidad de tres especies de palmas. Informe Anual, Programa de Diversificación, FHIA, La Lima, Honduras.
Tabora, P.C., Medlicott, A. Ramirez, T. and Salgado, T. (1991) Evaluation of palm heart sources in Honduras, in *Proceedings International Conference on Pijuayo,* Iquitos, Peru.
Tucker, R. (1988) *The Palms of Subequatorial Queensland,* Palm and Cycad Societies of Australia, Brisbane, Australia.
Uhl, N.W. and Dransfield, J. (1987) *Genera Palmarum,* Allen Press, Lawrence, Kansas.
Vargas, E. (1989) Enfermedades del tallo de palmito. *Serie Técnica Pejibaye,* UCR, Boletín Informativo, **I**, 12–13.
Vargas, A. (1991) Evaluación de ocho densidades de población en combinación con tres tallos por cepa en pejibaye para palmito (*Bactris gasipaes*), CORBANA, Informe Anual.

Velasco, A. (1989) En Colombia, el palmito de Chontaduro (*Bactris gasipaes*), *Serie Técnica Pejibaye*, UCR, Boletín Informativo, **I** (2).
Zamora, C. (1989) *El programa nacional de palmito de pejibaye.* ASBANA, San José, Costa Rica.
Zamora, C. (1990) *Algunos Aspectos Sobre la Situación del Palmito de Pejibaye en Costa Rica*, ASBANA, Costa Rica.
Zantua, M., Tabora, P. and Ramirez, T. (1989) Preliminary observation of some palm heart sources in Honduras, in *Proceedings of the Consultative Meeting on Palm Hearts*, FHIA, La Lima, Honduras, pp. 39–41.

8

Vegetable soybean (*Glycine*)

T. E. Carter, Jr. and S. Shanmugasundaram

The global importance of soybean (*Glycine max* (L.) Merill) has fostered a great body of research to improve production, management, and breeding of the crop. Much of the research in the West is summarized in the form of monographs edited by Caldwell (1973) and Wilcox (1987). Brar and Carter (1992) have reviewed recent advances in breeding for soyfoods and the application of biotechnology to soybean breeding. The purpose of this present account is to review the history and use of soybean as a green vegetable (green bean), a food not well known in the West but one of great popularity in the East. Known as *edamame* in Japan, *mao duo* in China, and *poot kong* in Korea (Lumpkin and Konovsky, 1991), the green vegetable is perhaps the tastiest of the traditional soyfoods.

Throughout much of Asia, the soybean is consumed as a delicious fresh vegetable in season. Over the centuries, specialized landraces have evolved for this specific food use and in the twentieth century modern vegetable-type cultivars have been developed from them in Japan, Taiwan, and the Peoples Republic of China (PRC). To emphasize the distinctiveness of vegetable versus grain soybean, the green vegetable soybean is hereafter referred to as *edamame* (pronounced 'aid ah mommy').

8.1 A BRIEF HISTORY OF SOYBEAN

It is commonly held that soybean was domesticated from its weedy relative, *Glycine soja* Sieb. and Zucc., also known as wild soybean (Hymowitz and Bernard, 1991). The wild soybean is found today along roadsides and riverbanks in the PRC, but the cultivated soybean is dependent upon mankind for survival, growing only in farmers' fields. Soybean was domesticated in eastern China as a food crop more than 3000 years ago. The crop then spread westward by caravan to present-day India and Nepal via the

Pulses and Vegetables. Edited by J. T. Williams. Published by Chapman & Hall, 2–6 Boundary Row, London SE1 8HN. ISBN 0 412 46610 4.

Silk Road, and eastward by sea-going vessels to Japan before the sixth century AD (Hymowitz and Bernard, 1991). Marco Polo made no reference to the soybean in his memoirs (*ca.* AD1300) and it appears that missionaries returning from China may have been the first to bring the soybean to Europe early in the eighteenth century (Hymowitz and Newell, 1981). The introduction of soybean into the USA is credited to an American sailor returning from the Orient in 1765 (Hymowitz and Harlan, 1983). Benjamin Franklin reintroduced the soybean to the USA some years later from Europe. Early in the twentieth century, plant explorers Dorsett and Morse returned from China with more than 4000 landraces (Hymowitz, 1984). Introduction of these landraces formed the foundation for modern-day soybean production in North and South America.

At the time of introduction to the West, soybean was a well established part of the human diet in Asia. The soyfoods recipes that made soybean popular in Asia did not accompany the crop to the West. Soybean was not recognized as a food crop in those early days, finding a niche in Western agriculture initially as a hay and green manure crop. Early in the twentieth century the future of soybean was forever changed (both in the East and West) by the discovery that soybean seed is a rich source of protein and oil (approximately 40 and 20% respectively). The economic importance of the discovery shifted the soybean rapidly from forage to seed production in North America, and the first commercial extraction of soybean oil commenced in Elizabeth City, North Carolina in 1915. The technology for extraction, adapted from cotton seed presses already in use, elevated soybean to the status of premier grain crop in the West, with production soon outstripping that in the East. Now grown on 50 million ha worldwide, more than 90% of the crop in the West and perhaps 30% in the East is devoted to oil extraction (Brar and Carter, 1992; Smith and Huyser, 1987; and USDA-FAS, 1991). Soybean production now accounts for 35% of the world's vegetable oil market. The protein meal remaining after oil extraction is equally valuable and used extensively as livestock feed.

8.2. FOOD USES

Because of the economic importance of soybean protein and oil in the twentieth century, it is easy to forget that in Asia the soybean remains a traditional part of the diet for over one billion people (Chen and Buckwell, 1991) as well as being prized as a source of cooking oil and livestock feed. It is the lengthy interaction of mankind with soyfoods, both the fresh vegetable and foods prepared from the mature bean, that may be responsible for the rich genetic diversity found in Asian soybean landraces today.

The array of morphological diversity is much greater in soybean than in the wild progenitor. Ancient farmers played an important role in developing the current reservoir of genetic diversity in the crop, by noticing and preserving novel genetic variants that would have been lost otherwise. The most genetically diverse trait in soybean is seed appearance (Nelson and Dieyang Wang, 1989), a trait that is strongly associated with specific food uses. Brar and Carter (1992) speculate that the association is so strong that evolution of soyfoods recipes may have been concurrent with domestication and diversification of the crop. Over long spans of time, mutants for seed colour, size, etc. may have been noticed and saved by early cultivators. These traits would eventually become associated with evolving soyfoods recipes of the particular village in which they first appeared. As trade developed, unique recipes and distinctive seed may have come to be exchanged as sets. Thus, an association between a particular food use and a visually identifiable seed type could have originated. The proliferation of hundreds of traditional soyfoods recipes and their preservation to modern times may have, by association, helped preserve the genetic diversity of the soybean. Today more than 15 000 distinct soybean landraces exist. This phenomenon, involving the eye and palate of mankind, is called 'selection for perceptual distinctiveness', and has been noted as a mechanism of preservation for diversity in cassava (*Manihot esculenta*) (Boster, 1985).

The relationship between seed appearance and intended food use in Asian soybean landraces is such that it is possible to examine the seed and make an educated guess regarding food use. For example, varieties with small flat black seeds (similar to wild soybean) are often used to make a spicy fermented garnish for vegetable dishes in southern China. Small, round, and yellow seeds are used to make a fermented product called natto in Japan, first described about AD1000. Larger yellow seeds are used to make tofu in Japan. Tofu, a low calorie white curd made from soybean, was described in Chinese writings more than 2000 years ago. Extremely large black seeds (> 50g/100 seed) are cooked and eaten as a New Year's dish in Japan to assure good health in the coming seasons (Watanabe and Kishi, 1984). The presence of a green cotyledon or seed coat in the mature seed usually indicates that the variety is intended for vegetable use. About one-half of the vegetable varieties in China exhibit the green cotyledon trait.

8.3 *EDAMAME*

Lumpkin and Konovsky (1991) postulate that green immature soybean may have been gathered and eaten for as long as man has cultivated the crop. The earliest reference to a vegetable soybean may be found in Chinese writings dating to the second century BC (Shurtleff and Aoyagi, 1989). Medicinal uses of the fresh green bean were cited in the fifth century AD (Shurtleff and Aoyagi, 1989) and even today the folk wisdom of medicinal benefit remains in China.

Edamame is the most easily prepared and tastiest of the soyfoods. Unlike tofu and natto which require several steps in processing, *edamame* is simply prepared by boiling the green immature seed, either in the pod or shelled, for 5–10 mins in salt water. The shelled bean is cooked and traditionally served in China as a component of a prepared dish. *Edamame* is often cooked in the pod and consumed as finger food with tea or beer in Japan; with a little practice the bean can be squeezed out of the boiled pod and directly into the mouth.

Fresh *edamame* is sold at market in Asia as pods attached to the plant, loose pods, or shelled beans. Although the pod and plant are inedible, it is widely believed in Japan that the keeping quality (sweetness and flavour) of the *edamame* is better when left in the pod and, if possible, attached to the plant until needed. Masuda (1991) indicates that sweetness, flavour, and aroma all begin to decline noticeably after 10 h storage at 26°C, requiring special handling for consumer acceptance. Tsay and Sheu (1991) note that the decline in fresh *edamame* quality can be slowed greatly by lowering the temperature to 5°C and storing the beans in plastic bags. *Edamame* is remarkably well suited to the fresh frozen market and, since 1975, has been the basis of a large and growing export industry in Taiwan. The principal market is Japan, with most *edamame* sold in the pod in 500 g plastic bags.

8.4 CONSUMPTION AND MARKETING OF *EDAMAME*

The three geographical areas most noted for *edamame* production are the PRC, Taiwan and Japan. In the PRC, most soybean production is for harvest of the mature seed. Although farmers will consume grain-type soybeans as a vegetable when in season, this is customary only for home use. Specialty landraces and cultivars for vegetable production are found typically only near urban centres or along railways for easy transport to fresh market. These soybean types often have a green seed coat at maturity and are thought to have a taste superior to the grain-type bean. Vegetable soybean is probably more popular in southern regions of the PRC than in the north, perhaps because of the extended growing season and greater opportunity for consumption. Even so, less than 2% of the hectarage in Jiangsu province in southern PRC is devoted to vegetable-type varieties (Prof. Gai Junyi, Nanjing Agricultural University, pers. comm.).

In Japan, consumer demand for *edamame* rose sharply from 1965 to 1985 and then remained steady for the next five years (Nakano, 1991 and Table 8.1). Although domestic production of fresh *edamame* has increased with market expansion, more than 20% of the *edamame* consumed in Japan is now imported from Taiwan as frozen pods. Given the profitable *edamame* market, one might expect that frozen *edamame* would be produced in Japan for consumption throughout the year. Such is not the case, however. The likely explanation is that *edamame* is competing with many other fresh vegetables for scarce agricultural land (Nakano, 1991). Frozen *edamame* does not command the same high price as fresh *edamame*; it is an economic

Table 8.1 *Edamame* production and consumption in Japan

Year	*Area (1000 ha)*	*Production (1000 t)*	*Import (1000 t)*
1965	6.4	61	<1
1970	9.4	91	<1
1975	10.8	99	6
1980	14.1	119	26
1985	14.3	116	31
1990	–	120	36

Note: Japan produced 400 000 t of grain soybean in 1989.
t, metric tonne.
Source: Japanese Ministry of Agriculture, Forestry and Fisheries, cited by Lumpkin and Konovsky (1991).

consequence that frozen *edamame* is more cheaply exported from Taiwan than produced in Japan.

In Taiwan, most of the soybean hectarage is devoted to *edamame,* with production concentrated in the south end of the island. In this tropical environment, *edamame* can be produced as a spring, autumn, or to a small degree, a summer crop. (Tsay *et al.*, 1991). About 80% of the crop is exported to Japan as fresh frozen *edamame* with the balance consumed locally (Table 8.2). The crop is not grown on a large scale in the summer months because heavy rainfall and high temperatures are more conducive to rice than to soybean production. Also, Japanese farmers situated in more northerly latitudes meet the needs of the Japanese *edamame* market very well in the summer months, so that demand and price for frozen *edamame* from Taiwan are both relatively low.

8.5 HARVEST FACTORS

Edamame is harvested as a plump green pod just prior to the yellowing and dehydration that accompany senescence. In the nomenclature of Fehr *et al.* (1971), *edamame* is harvested between R6 and R7 growth stages. Although

Table 8.2 *Edamame* production in and export from Taiwan

		Area (ha)		*Production (mt)*			
Year	*Yield kg ha^{-1}*	*Spring*	*Autumn*	*Spring*	*Autumn*	*Export (mt)*	*Export value US$/kg*
1984	5777	2577	4092	13 308	25 220	33 818	–
1986	5832	3116	5349	19 611	29 761	36 321	1.12[a]
1988	6091	3896	3751	23 017	23 554	36 321	1.31
1990	6411	5092	4760	32 500	30 663	37 055	2.04

[a] 1987 value. Source: Cheng, (1991).

Table 8.3 Relation between days after flowering and *edamame* quality. The grading is for the cultivar Sapporomidori, grown in Japan

	Days after flowering		
Quality Trait	*30*	*40*	*50*
Taste			
Free amino acids	Good	Normal	Bad
Sugars	Bad	Good	Good
Pod colour	Good	Good	Bad
Degree of pod expansion	Bad	Normal	Good

Source: Chiba, (1991).

flowering and setting of pods may extend over a two or three-week time period, the extreme sensitivity of soybean to photoperiod tends to synchronize filling and yellowing of pods on a plant so conveniently that only one harvest per crop is necessary.

Timing of harvest is critical for high quality *edamame*, yet often difficult to determine. The yellowing of the pod, signifying a late harvest, represents a decline or degradation of free amino acids in the seed, a constituent associated with good taste. Premature harvest results in pods that are too flat and with less sweetness than normally associated with *edamame*, reducing yield and consumer acceptance simultaneously. Chiba (1991) has shown that optimum harvest time may involve a narrow window of 10 days or less (Table 8.3). Correct timing of harvest is often dependent upon the grower's familiarity with a particular variety and the growing conditions.

The person hours required to grow and manually harvest a hectare has been estimated as 2000 and 700 in Japan and Taiwan, respectively (Nakano, 1991; Hsieh and Su, 1991). Because of the high labour investment in the crop, recent advances in technology, including stripper harvesters and pod sorting devices, are being adapted to mechanize the harvest in Taiwan (Hsieh and Su, 1991).

8.6 *EDAMAME* PRODUCTION FACTORS

8.6.1 **Cropping system**

The soybean is an erect, annual, self-pollinated legume that is highly sensitive to photoperiod. Cultivation practices for *edamame* are similar to those for grain-type soybeans and are reviewed by Hung *et al.* (1991), Johnson (1987), Kokobun (1991), Pandey (1987), and Tsay *et al.* (1991). In latitudes greater than 36° (absolute value), where the growing season is short, only one crop of soybean is grown per year with planting and grain harvest in May and September, respectively. In latitudes between 27° and 36°, as many as three different cropping systems may be employed. In the

USA, full-season and late-season plantings (commonly defined as planting either before or after June 15) are useful in crop rotations. Grain harvest is in October or November. In the PRC, three planting systems are employed. Spring soybean is planted in April or May and harvested in June–July. Summer soybean is planted in June and harvested in August. Autumn soybean is planted in August and harvested in October. For latitudes less than 25°, soybean has the potential to be planted year round with actual cropping systems dependent upon local rainfall and temperature conditions. In southern Taiwan, spring soybean is planted and harvested in February and June, respectively. Autumn soybean is planted in October and harvested in January (Tsay *et al.*, 1991).

In Japan, *edamame* can be produced in either an early maturity, normal season, or late cropping production system (i.e. three different planting times) in a rotation with other vegetables (Kokobun, 1991). In some areas of the country, fresh edamame is grown in heated glasshouses in the winter, or transplanted to unheated vinyl tunnels early in the spring (known as 'forcing' and 'semiforcing'), so that the crop can be harvested out of season and command a high price (Kokobun, 1991). In the southern PRC, rice is the predominant crop. Therefore, when vegetable soybean is sown as a spring crop, it will most likely be planted in an upland field unsuited to paddy rice production. Fall vegetable soybean is often sown after rice is harvested, while summer soybean is often sown on the dikes of the rice paddies. In Taiwan, soybean is sown primarily in the spring or fall in rotation with the predominant crop, paddy rice.

8.6.2 **Photoperiod sensitivity**

To fit soybean into a cropping system, careful cultivar selection must be exercised with the single most important attribute being its photoperiod response. A cultivar suited, by photoperiod response, for spring planting may not fare well in a fall planting at the same location or in a spring planting 400 km north or south. The impact of photoperiodic response is so great that when planted on the same day, some varieties may mature just as others begin to flower. At least four genes are known to control photoperiod sensitivity with soybean cultivars differing widely in the critical length of darkness required to induce flowering (Palmer and Kilen, 1987). In general, most varieties tend to lengthen the time span from planting to flowering as they are grown further from the equator. Days from planting to harvest can range from 80 to 150 depending upon genotype and environment. Cooler temperatures also tend to lengthen the time from planting to flowering.

Although photoperiod response is the single most important factor controlling reproduction in soybean, there is no universally agreed-upon system to classify photoperiod response. In the USA, cultivars are grouped according to photoperiod sensitivity into classes running consecutively from 0 to 10. Genotypes of group 0 are generally adapted to Canada, those of group 5 to Virginia and North Carolina latitudes, and group 10 to the equator. In the PRC, one must know the province of origin and planting type (spring, summer, or fall) to estimate the photoperiodic response of a cultivar. In Japan, at least nine photoperiod ecotypes are recognized by breeders, although designations of early, medium, and late maturity are in common usage in the seed trade. In latitudes greater than 37° (absolute value), cultivars planted at similar latitudes in Asia and the West tend to have similar photoperiodic response. Below 37°, because of diverging cropping patterns, Asian cultivars tend to require a longer dark period to induce flowering than do Western cultivars grown at similar latitudes. As a result, when planted on the same day, the Asian cultivars tend to mature two or three weeks earlier than the Western cultivars.

In attempting to move *edamame* production to new areas such as Latin America or Africa, it is best to begin with field trials that include *edamame* cultivars from Asia that perform well at a similar latitude and climate. Comparisons with locally adapted grain-type soybean cultivars is important because specific disease problems and soil conditions of a region may not be conducive to production of introduced *edamame* cultivars. Locally adapted grain-types may find acceptance as a vegetable when there is no preconceived notion of taste by the consumer. Consumer tastes are known to vary within Japan (Nakano, 1991) and from country to country.

8.7 NUTRITIONAL CHARACTER OF *EDAMAME*

Edamame, like the mature seed, is rich in protein, unsaturated fats, phosphorus, iron, thiamine, riboflavin, and is a fair source of calcium (Tables 8.4 and 8.5). It is higher in protein, oil, and calcium than the vegetable green pea (*Pisum sativum*). The level of saturated fat in the oil of *edamame* is about the same as for the mature soybean while the level of oleic (18:1) acid is higher and linoleic (18:2) acid lower. No health risks have been attributed to consumption of 18:1 although there is some concern that high consumption of 18:2 could result in undesirable side-effects such as hypertension (Shukla, 1990). The oil composition of *edamame* is thus marginally better than for the mature bean.

Soy protein has been thought to be of poorer quality than animal protein with respect to human nutrition, principally because of inadequate levels of the sulphur-containing amino acids, cystine and methionine. Recent findings reviewed by Young (1991), however, indicate that soy protein is of high nutritional value and that the concentration of sulphur-containing amino acids is entirely sufficient for adults. Early estimates of amino acid requirements for humans were developed using species other than *Homo sapiens*, thus seriously underestimating the requirements for essential

Table 8.4 Seed composition of *edamame* (vegetable soybean), green pea (*Pisum sativum*) and mature soybean seed

	Edamame			*Mature soybean*	*Green pea*		
Nutrients and Units[a]	*FAO*[c]	*USDA*[d]	*Japan*[f]	*USDA*[e]	*FAO*[c]	*USDA*[d]	*Japan*[f]
Food energy (kcal)	437	452	429	454	385	383	395
Protein (g)	40.8	39.9	39.4	43.7	25.4	25.6	30.0
Total lipid (g)	17.9	20.9	22.8	21.8	1.6	1.9	0.8
Total carbohydrate (g)	35.8	34.0	32.2	29.2	69.2	68.4	65.4
Crude fibre (g)	6.0	6.3	6.6	5.4	9.8	10.5	11.9
Ash (g)	5.3	5.2	5.5	5.3	3.7	4.1	3.7
Calcium (mg)	245	606	242	303	131	118	115
Iron (mg)	11.9	10.9	5.9	17.2	4.9	7.0	7.8
Magnesium (mg)				306		156	
Phosphorus (mg)	496	596	484	770	418	511	288
Potassium (mg)			484	1965		1154	288
Sodium (mg)			3	2		23	12
Zinc (mg)				5.35		5.86	
Copper (mg)				1.842		0.833	
Manganese (mg)				2.752		1.939	
Ascorbic Acid (mg)	84.9	89.2	93.4	6.6	110.0	189.2	74.0
Thiamin (mg)	1.25	1.34	0.93	0.96	1.15	1.26	1.02
Riboflavin (mg)	0.53	0.54	0.48	0.95	0.45	0.62	0.08
Niacin (mg)		5.08	3.5	1.78		9.89	7.8
Vitamin A[b]	1132	554	346	26	1659	3072	1481

[a] Amount in 100 g of seed expressed on a 0% moisture basis.
[b] USDA units expressed as IU, other sources mg.
[c] FAO (1972).
[d] Haytowitz and Matthews (1984).
[e] Haytowitz and Matthews (1986).
[f] Standard Tables of Food Composition in Japan, fourth edn. (1982).

Table 8.5 Fatty acid composition of oil from developing and mature soybean seed

Days after flowering[a]	*Fatty acids (mol %)* 16:0	18:0	18:1	18:2	18:3	*Saturated fat (%)*
N78-2245[b]						
30	13.1	3.9	54.1	23.7	5.2	17.0
45	10.8	3.1	63.7	18.1	3.9	13.9
60	10.8	4.4	55.7	25.0	3.8	15.2
75	10.9	4.5	47.2	32.9	4.5	15.4
Dare[c]						
30	15.1	3.2	25.0	47.4	9.3	18.3
45	12.7	3.4	25.0	51.3	7.2	16.1
60	12.8	3.3	21.4	54.9	6.9	16.1
75	12.2	3.1	21.0	56.0	7.1	15.3
Japanese breeding material[d]						
25	16.3	3.5	35.8	32.1	11.4	19.8
40	11.1	2.3	51.0	29.4	6.2	13.4
45	8.7	1.9	57.2	27.6	4.6	10.6
50	9.2	1.8	42.5	41.1	5.5	11.0
75	10.0	2.3	21.9	58.2	7.5	12.3

[a] *Edamame* normally harvested between 35 and 45 days after flowering.
[b] A low 18:3 germplasm release from USDA-ARS, Carver *et al.*, (1984).
[c] A cultivar with oil composition typical for the southeastern USA, Carver *et al.* (1984).
[d] Kajimoto *et al.* (1982) and cited by Masuda (1991).

amino acids (by a factor of two), except for cystine and methionine (Young, 1991). Using revised estimates for amino acid requirements and the amino acid composition values in Table 8.6, the limiting amino acid(s) for a protein source can be identified in the following way: for each essential amino acid divide the figure in column 8 by the corresponding amino acid composition(s). (Cystine–methionine and phenylalanine–tyrosine requirements are given as a sum for the pairs in Table 8.6.) The highest value for a particular protein source indicates the most limiting amino acid. Following this process, one sees that the protein of the mature soybean seed is very well balanced. The protein of *edamame* is only slightly less balanced, being fractionally lower in cystine + methionine concentration than protein of the mature bean. The protein of the green pea is less balanced than that of either *edamame* or the mature bean because it is deficient in tryptophan. Based upon a 'most limiting factor' basis, 1 g of protein from the mature soybean is equivalent to 1.16 g of *edamame* protein or 1.29 g of green pea protein in the diet of a healthy adult. Because *edamame* is one-third higher in protein than the green pea and more balanced in protein composition, *edamame* must be regarded as an excellent vegetable protein source.

Table 8.6. Amino acid composition of seed protein from *edamame* (vegetable soybean), green pea (*Pisum sativum*) and mature soybean seed (g/100 g protein)

Amino Acids	Edamame		Mature soybean			Green pea	
	USDA	*Yazdi-Samadi*[a] et al. *(1977)*	*USDA*[d]	*Yazdi-Samadi*[a] et al. *(1977)*	*Hafez*[b] *(1983)*	*USDA*[c]	*Adult human requirement*[f] *(mg/kg body wt./day)*
Trytophan[e]	1.24	–	1.37	–	1.53	0.86	6
Threonine[e]	4.09	4.4	4.11	4.9	3.01	4.73	15
Isoleucine[e]	4.51	4.6	4.59	4.7	3.59	4.55	23
Leucine[e]	7.33	7.4	7.70	7.5	6.23	7.53	39
Lysine[e]	6.13	6.1	6.29	6.5	5.14	7.39	30
Methionine[e]	1.24	1.2	1.27	1.2	0.70	1.91	
Cystine[e]	0.93	0.9	1.52	0.9	1.92	0.75	(Σ=15)
Phenylalanine[e]	4.64	4.8	4.93	5.2	3.67	4.66	
Tyrosine[e]	3.67	4.1	3.58	4.1	4.42	2.66	(Σ=39)
Valine[e]	4.56	4.6	4.71	4.9	3.64	5.48	20
Arginine	8.25	8.4	7.33	8.7	6.44	9.98	
Histidine	2.76	2.6	2.55	2.7	2.19	2.50	
Alanine	4.61	5.4	4.55	4.7	3.37	5.60	
Aspartic Acid	11.95	13.1	11.89	13.1	9.27	11.57	
Glutamic Acid	19.27	18.4	18.31	20.4	14.69	17.28	
Glycine	4.27	4.2	4.37	4.4	3.46	4.29	
Proline	4.81	4.3	5.53	4.6	–	4.03	
Serine	5.71	5.8	5.47	6.0	4.23	4.22	

[a] Average for two midwestern USA varieties. Composition estimated for soybean seed at 35 days after flowering, the approproximate harvest date for *edamame*, and at maturity.
[b] Average for 14 midwestern USA varieties.
[c] Haytowitz and Mathews (1984).
[d] Haytowitz and Mathews (1985).
[e] Essential amino acids for man.
[f] Revised amino acids requirements, Young (1991).

8.8 PREFERENCE FACTORS FOR *EDAMAME*

Acceptance of *edamame* by consumers in Japan is governed by appearance, sweetness, flavour, aroma, and texture. Bright green pods sell most quickly while yellow pods are unacceptable because they usually indicate a decline in taste and in vitamins such as vitamin C (Masuda, 1991). Two-seeded and three-seeded pods are preferred over one-seeded pods because it is easier to squeeze the seed from the pod after boiling. In Japan, very large seed (>30 g/100 seed at maturity) may be preferred but medium size seed (25 g/100 seed at maturity) is accepted. Sweetness is primarily a function of sucrose content which increases as the *edamame* nears the optimum harvest date (Table 8.3). When the pods begin to senesce, sucrose may decline as the oligosaccharides, stachyose and raffinose, accumulate to relatively high levels (Table 8.7). Although raffinosaccharides are thought to be partially responsible for the flatulence associated with cooked beans (Brar and Carter, 1992), these sugars are not concentrated in *edamame*. Masuda (1991) indicates that *edamame* genotypes tend to be higher in sucrose than grain-type cultivars.

The level of free amino acids may play a role in sweetness and flavour of *edamame* (Masuda, 1991; Tsou and Hong, 1991). Free amino acid concentration decreases as seeds approach maturity, explaining in part the associated decline in flavour (Chiba, 1991). Glutamic acid, alanine, and asparagine are present in highest concentration as free amino acids in *edamame* (Chiba, 1991). The level of free amino acid concentration also declines in post-harvest storage, again associated with an overall decline in flavour and sweetness (Chiba, 1991).

Table 8.7 Soluble sugar content in soybean and seed harvested at 30–40 days after flowering and at maturity (mg/g dry wt.)

Soluble sugar	*Tsou and Hong*[a] *(1991)*	*Saravitz* et al. *(1987)* HST[b]	*Saravitz* et al. *(1987)* LST[c]	*Yazdi-Samadi*[d] et al. *(1977)*
30–40 days after flowering[e]				
Sucrose	91	54	56	56
Stachyose	1	<1	<1	<1
Raffinose	<1	<1	<1	<1
Maturity				
Sucrose	62	33	55	49
Stachyose	25	42	29	24
Raffinose	15	4	10	2

[a] Mean for three soybean genotypes grown for *edamame* in Taiwan.
[b] Mean for two soybean genotypes with high stachyose content in mature seed.
[c] Mean for two soybean genotypes with relatively low stachyose content in mature seed.
[d] Mean for a midwestern USA cultivar and Japanese *edamame* cultivar.
[e] Approximate harvest date for *edamame*.

The beany flavour of soybean is associated with hexanal compounds that result from the degradation of fatty acids. Although some consumers are accustomed to, and prefer, the beany flavour of *edamame*, others prefer a milder flavour. *Edamame* and the dry bean vary in their concentration of hexanal compounds as do the different cultivars. Production of hexanal compounds is controlled in part by lipoxygenase enzymes (Davies *et al.*, 1987). Soybean has three forms of the enzyme for which null variants have been found (Davies and Nielsen, 1987). Many *edamame* cultivars either lack or have reduced activity in the L1 lipoxygenase enzyme. The gene for this trait is being used in the breeding of *edamame* cultivars with mild flavour. Further details can be obtained from the review by Brar and Carter (1992).

Many other compounds affect the flavour and aroma of *edamame*. Masuda (1991) cites 11 additional compounds that negatively affect, and one that positively affects, the flavour or aroma of *edamame*. Because the research is relatively new, the impact of these findings on production and breeding of *edamame* is not clear at present. It is expected that new breeding strategies for *edamame* will result.

Seed texture is also important in *edamame*. Prolonged storage or late harvest will cause *edamame* to develop a hard texture that is unappetizing (Chiba, 1991; Tsou and Hong, 1991). The Japanese landrace Tanbaguro is reported to have a moderate texture compared with normal soybean cultivars (Masuda, 1991).

8.9 BREEDING *EDAMAME* CULTIVARS

8.9.1 **Genetic base**

Breeders have a wide genetic base from which to select suitable *edamame* characteristics. More than 400 landraces of *edamame* exist in Japan (Lumpkin and Konovsky, 1991). Approximately 15 000 landraces of all types are present in the PRC soybean germplasm collection (Prof. Gai Junyi, Nanjing Agricultural University, pers. comm.). Because of the frequency of green seed coat types, it is probable that several hundred vegetable accessions exist in this collection. Approximately 150 *edamame* landraces from various parts of Asia are maintained at the Asian Vegetable Research and Development Centre in Taiwan, China. In the USA, the consortium of 1890s Historically Black Colleges is assembling and evaluating a collection of vegetable and other food use soybean accessions (M.Rangappa, Va. State University, pers. comm.). Doubtless, many *edamame* accessions exist in the 13 000 accession USDA soybean germplasm collection (a part of the NPGS) at Urbana, Illinois (Juvik *et al.*, 1985) although they are not usually identified as such.

8.9.2 **Breeding programmes**

The breeding of soyfood cultivars in the USA is reviewed by Brar and Carter (1992); Wilcox (1990) and Carter (1988). The cultivars Emerald and Verde were released for home garden use (Bernard *et al.*, 1988) and a few small breeding programmes are underway to develop *edamame* cultivars for production near urban centres with large Oriental populations. No popular cultivars have been released, however. Several large-seeded cultivars have been released for production as an export crop to meet the demands of the Japanese tofu market. Some of these cultivars may have cross-over appeal for *edamame* production in the USA, but prospects are unclear at present (Brar and Carter, 1992).

Edamame breeding efforts in Taiwan are reviewed in detail by Shanmugasundaram *et al.* (1989, 1991). Large-scale evaluation of Japanese *edamame* cultivars and landraces for use in Taiwanese production began in the 1980s, with Tsuronoko and Ryokkoh identified as superior types for spring and fall sowing. More recently a selection from Taisho Shiroge (a Japanese *edamame* cultivar) has been released as Kaohsiung No. 1 and occupies 84% of the Taiwanese *edamame* hectarage. A number of promising parents have been identified from cultivar evaluation studies. These lines are being hybridized to develop new, more productive progeny with greater consumer acceptance in Japan. The development of criteria for progeny selection is an evolving process. Current criteria for the ideal *edamame* genotype for Taiwan is found in Table 8.8.

A partial listing of important *edamame* cultivars currently in use in Japan is found in Table 8.9. More than 50 cultivars have been developed by private seed companies in Japan in the past 30 years. Even so, some popular landraces and publicly released materials are grown for this market. Pedigrees of some of the more important cultivars are described by Lumpkin and Konovsky (1991). Camacho (1989) and Takahashi (1991) provide further descriptions of some of the most important *edamame* cultivars.

In the PRC, breeding efforts are centred on improvement of grain type soybean with little effort devoted to vegetable cultivars. Recently, Dr Song at the Jiangsu Agricultural University in Nanjing has released two modern *edamame* cultivars, Nan Nong 87C-38 and Ning Qing Dou No. 1. Both have green seed coats, a mature 100-seed weight of 25 and 35 g, respectively, and protein content about 48%. These lines are the progeny of a Chinese landrace for vegetable production hybridized to high yielding breeding lines developed at Nanjing. In Liaoning province, Professor Hechong has released three vegetable cultivars: Dan dou Nos. 4, 5, and 6.

8.9.3 **Breeding strategy**

Developing a breeding strategy for *edamame* cultivars is not simple because

Table 8.8 A partial listing of ideal characteristics for *edamame* cultivars in Taiwan

Character	*Range*
Pod appearance	
Pod and seed colour	bright green, free from blemishes
Pod length	≥ 5.0 cm
Pod width	≥ 1.4 cm
Frequency of 1-seeded pod	< 25%
Pubescence density	?
Plant appearance	
Leaf	large, lanceolate (promotes good pod colour and reduces 1-seed pod frequency)
Pubescence, colour	grey
Branching	minimal
Pod clearance	at least 10 cm from soil
No. of nodes on main stem at harvest	10–14
Flowering	≥ 40 days after planting
Lodging	minimal
Seed appearance	
Hilum colour	grey or light brown
100-seed weight	≥ 30g
Harvest	pods easy to strip, longer duration from R6 to R7 (the harvest window)
Resistance to	downy mildew, bacterial pustule, soybean rust, high temperature

Source: Shanmugasundaram *et al.* (1991).

of the potentially large number of traits that may fall under selection. In addition to the usual agronomic traits involved in the breeding of grain-type soybeans, seed and pod size, taste, aroma, sweetness, texture, and colour may also be considered. Harvest is more labour intensive for the green bean than the mature seed so that yield data are more difficult to generate. The concentration of certain seed chemical components of the green bean can be difficult to measure, change diurnally, or degrade rapidly after sampling so that accurate comparison of a large number of genotypes, required in a breeding programme is not possible. Because of the difficulty in defining consumer acceptance biochemically, sensory evaluation panels are often used to compare elite breeding lines after progeny have been highly selected for morphological and agronomic traits. More sophisticated analytical techniques may be reserved for characterizing the potential parents for a breeding programme rather than progeny.

The more important morphological characters to be selected are listed

Table 8.9 A partial listing of soybean cultivars used for *edamame* production in Japan and Taiwan

Cultivar	*Origin*	*Maturity*	*Notes*[a]
Datacha	Landrace	Very late	Brown seed, sweet taste
Hiradoko	Landrace	Very late	Green flat seed
Krosai-Chamame	Landrace	Very late	Brown seed
Kinehu	Landrace	Late	Often used as a standard for comparison
Mikawashima	Landrace	Medium	
Tanbaguro	Landrace	Very late	Extremely large black seed, moderate texture
Okuhara 1 Gou	Public	Very early	Yellow seed
Wase Midori	Public	Medium	Green seed
Tokyo Wase	Public	Late	Yellow seed
Iwa-Mame-Kei-1	Public	Late	Green seed
Iwa-Mame-Kei-4	Public	Medium–late	Light-green seed
Kachsiung No. 1[b]	Public	Medium	Currently most popular cultivar in Taiwan, selection from Japanese cultivar Taisho shiroge
Echigo Musume	Private Co.	Late	Green seed
Fukura	Private Co.	Early	Often used as standard for comparison
Gokuwaseozaya	?	Very early	Glass house and early season production
Green 75	?	Very early	Very large seed
Hakuchou	Private?	Very early	Glasshouse and early season production
Hatu Musume	Private	Late	Light green seed
Karicachi	Private	Medium–early	
Kogen Midori	Private?	Medium	
Kita no Siki	Private	Early–medium	Light green seed
Kitanosuzu	Private	?	Tolerant to temperature stress
Mikawajima	?	Medium	
Mosono Green	Private	Early–medium	
Okubarawase	?	Very early	Glasshouse and early season production
Sapporo Midori	Private	Early	Most popular for fresh market
Sayamusume	Private	Early–medium	Possibly long harvest period
Shiratsuyu	?	Medium	
Ryokkoh[b]	Private	Medium	Known as 305 in Taiwan
Tamasudar	?	Early	
Tengamine	Private	?	High sucrose and glutamic acid content
Tsuronoko[b]	Private	Medium	Dark green pods good for freezing, known as 205 in Taiwan
Wase Shiroge	?	Early–medium	
Yukimusume	Private	Early–medium	

[a] Source: Shanmugasundaram *et al.* (1991), Takahashi (1991), Lumpkin and Konovsky (1991), Masuda (1991).
[b] Leading *edamame* cultivars in Taiwan.

in Table 8.8. Other goals may involve improvement of consumer acceptance through selection of breeding lines having green beans with high sucrose and free amino acid content, low levels of hexanal [and derivatives], reduced isoflavin and saponin content, and soft texture. Nutritional objectives might include lowering the saturated fatty acid of the oil to below 10% and increasing the protein content of the seed to above 45% (on a dry weight basis).

Since 1945, a global effort has been underway to improve the disease and insect resistance, yielding ability, and seed composition of grain soybean cultivars. These efforts have resulted in the release of over 400 modern cultivars in the PRC (Prof. Gai Junyi, Nanjing Agricultural University, pers. comm.), 100 public cultivars in Japan (Dr Shoji Miyasaki, National Institute of Agrobiological Resources, pers. comm.), 100 in Latin America, and 290 public cultivars in the USA (Bernard *et al.*, 1988). These cultivars represent a reservoir of agronomic genes that should be transferred to new *edamame* cultivars. A great body of literature has accumulated describing the composition and yielding ability of these soybean genotypes with respect to the mature seed. Similar evaluation data for the green bean stage are relatively rare and furthermore difficult to obtain because of harvest difficulty and the necessity to store fresh samples prior to evaluation. It is suggested that for some traits, accumulated data on the mature plant may serve as an indirect measure of genetic potential of cultivars as parents in *edamame* breeding programmes. Cultivar rankings for mature and green seed are highly associated for the trait's seed yield, seed size, seed protein percentage, and amino acid and fatty acid composition so long as genotypes under comparison are of similar maturity (Carter *et al.*, 1982; Yazdi Samadi *et al.*, 1977, and Table 8.5). Descriptive information for cultivars at maturity should be directly applicable to the selection of parents in an *edamame* breeding programme. Absolute sucrose levels and perhaps genotypic rankings for this trait change drastically during maturation so that analysis of sugar content at maturity may reveal little about the sweetness of the green bean (Saravitz *et al.*, 1987 and Yazdi-Samadi *et al.*, 1977).

8.10 FUTURE PROSPECTS

The popularity of vegetable soybean remains high and may be increasing in Asia. Consumption has more than doubled in Taiwan in the last five years, and other Southeast Asian countries have begun to show some interest in commercial production as a result of contact with the Asian Vegetable Research and Development Centre (Shanmugasundaram *et al.*, 1991). The potential for the introduction of vegetable soybean to Latin America and Africa seems great. In Brazil, a breeding programme for *edamame* cultivars is underway (Carrio-Panizzi, 1989). At the head waters of the Amazon River, an area where protein and vegetables are in short supply, soybean has relatively few diseases and may be easy to produce as a high protein vegetable crop. The added value of soybean as a nitrogen fixer means that the vegetable is relatively economical to produce. Soybean production technology already exists in both Latin America and Africa so that the transition to *edamame* may involve only choice of cultivar and education regarding use as a vegetable.

In the USA, vegetable soybean is not commonly found on the grocer's shelf although it can be obtained from Oriental markets, as an import from Taiwan. The prospects for consumer acceptance may be good, however. Ten years ago, tofu was uncommon on grocers' shelves but now it is found in many major food chain stores. The entrance of this soyfoods product into the USA market has been facilitated by the rising health concern in the USA regarding cardiovascular diseases. Tofu is now seen as a healthy, low calorie, low saturated fat food. *Edamame* may gain acceptance in the USA because of the same health concerns. The route may be somewhat easier for this product than with tofu because it is tastier, involves little preparation, and is an easily consumed finger food.

Production of *edamame* is beginning to shift to low pesticide input in Taiwan and Japan because of public concern with safety issues. Because soybean is relatively disease and insect free compared to many other vegetables, the potential to produce *edamame* organically, especially in latitudes greater than 36° (absolute value) seems great.

REFERENCES

Bernard, R.L., Juvick, G.A. Hartwig, E.E. and Edwards, C.E. (1988). Origin and pedigrees of public soybean varieties in the United States and Canada. *U.S. Department of Agriculture, ARS Technical Bulletin* 1746, US Govt. Print. Office, Washington, DC.

Boster, J.S. (1985) Selection for perceptual distinctiveness: evidence from aguaruna cultivars of *Manihot esculenta. Econ. Bot.*, **39**, 310–25.

Brar, G.S. and Carter, T.E., Jr. (1992) The soybean breeding for soyfoods and biotechnology, in *Genetic Improvement of Vegetable Crops*, (eds G. Kalloo and B. Bergh), Pergamon Press, Oxford, UK (in press).

Caldwell, B.E. (ed.) (1973) *Soybeans: Improvement, Production and Uses*, American Society of Agronomy, Madison, WI.

Camacho-C., F. (1989) *Caracteristicas y uso potential para frijol de soja verde* (Edamame) *de las variedades de soja neueva y antiquas de la parte central de Japan.* Centro Para Investigaciones en Granos y Semillas, Universidad de Cost Rica San José.

Carrio-Panizzi, M.C. (1989) Vegetable soybeans in the East, in *Proceedings of World Soybean Research Conference IV*, (ed. A.J.Pascale), Buenos Aires, Argentina, 5 March–9 March 1989, Asociacion Argentina de la Soja, Buenos Aires, pp. 1101–5.

Carter, T.E., Jr. (1988) Genetic alteration of soybean seed size: Breeding strategies and market potential, in *Proceedings of American Seed Trade Association*, (ed. D. Wilerson) Washington, DC, **17**, 33–45.

Carter, T.E., Jr., Burton, J.W. and Brim, C.A. (1982) Recurrent selection for percent protein in soybean seed – Indirect effects on plant N accumulation and distribution. *Crop Science*, **22**, 513–19.

Carver, B.F., Wilson, R.F. and Burton, J.W. (1984) Developmental changes in acyl-composition of soybean seed selected for high oleic acid concentration. *Crop Science*, **24**, 1016–19.

Chen, Liang Yu and Buckwell, A. (1991) Chinese grain and economic policy, CAB International, Redwood Press Ltd., Melksham, UK.

Cheng, S.H. (1991) Vegetable soybean area, production, and foreign and domestic trade in Taiwan, in *Vegetable Soybean: Research Needs for Production and Quality Improvement*, (ed. S. Shanmugasundaram), Proceedings of a workshop held at Kenting, Taiwan, 29 April–2 May 1991, Asian Vegetable Research and Development Center, pp. 17–21.

Chiba, Y. (1991) Postharvest processing, marketing and quality degradation in vegetable soybean in Japan, in *Vegetable Soybean: Research Needs for Production and Quality Improvement*, (ed. S. Shanmugasundaram), Proceedings of a Workshop held at Kenting, Taiwan, 29 April – 2 May 1991. Asian Vegetable Research and Development Center, pp. 108–112.

Davies, C.S. and Nielsen, N.C. (1987) Registration of soybean germplasm that lacks lipoxygenase isozymes. *Crop Science*, **27**, 370.

Davies, C.S., Nielsen, S.S. and Nielsen, N.C. (1987) Flavor improvement of soybean preparations by genetic removal of lipoxygenase-2. *American Oil Chemical Society*, **64**, 1428–35.

FAO (1972) *Food Composition Table for Use in East Asia*, Food policy and nutrition division, FAO, Rome.

Fehr, W.R., Caviness, C.E., Burmood, D.T. and Pennington, J.S. (1971) Stages of development descriptions for soybeans, *Glycine max* (L.) Merrill. *Crop Science*, **11**, 929–31.

Haytowitz, D.B. and Mathews, R.H. (1984) *Composition of Foods: Vegetables and Vegetable Products*, U.S. Department of Agriculture, Handbook no. 8–11.

Haytowitz, D.B. and Mathews, R.H. (1986) *Composition of Foods: Legumes and Legume Products*, U.S. Department of Agriculture Handbook no. 8–16.

Hsieh, C.C. and Su, C.S. (1991) Management inputs and mechanical harvesting for vegetable soybean in Taiwan, in *Vegetable Soybean: Research Needs for Production and Quality Improvement*, (ed. S. Shanmugasundaram), Proceedings of a Workshop held at Kenting, Taiwan, 29 April – 2 May 1991, Asian Vegetable Research and Development Center, pp. 61–4.

Hung, A.T., Cheng, J.H., Ma, C.H. and Nobayashi, H. (1991) Effect of fertilizer management and *rhizobia* inoculation on yield and quality of vegetable soybean, in *Vegetable Soybean: Research Needs for Production and Quality Improvement*, (ed. S. Shanmugasundaram), Proceedings of a Workshop held at Kenting, Taiwan, 29 April – 2 May 1991, Asian Vegetable Research and Development Center, pp. 73–84.

Hymowitz, T. (1984) Dorsett-Morse soybean collection trip to East Asia: 50 year retrospective. *Economic Botany*, **38**, 378–88.

Hymowitz, T. and Bernard, R.L. (1991) Origin of the soybean and germplasm introduction and development in North America, in *Use of Plant Introductions in Cultivar Development, Part I*, (eds H.L. Shands and L.E. Wiesner), Crop Science Society of America, Madison, WI.

Hymowitz, T. and Harlan, J.R. (1983) Introduction of soybean to North America by Samuel Bowen in 1765. *Economic Botany*, **37**, 371–9.

Hymowitz, T. and Newell, C.A. (1981) Taxonomy of the genus *Glycine*: domestication and uses of Soybeans. *Economic Botany*, **35**, 272–88.

Johnson, R.R. (1987) Crop Management, in *Soybeans: Improvement Production and Uses,* (ed. J.R. Wilcox), American Society of Agron., Madison, WI.

Juvik, G.A., Bernard, R.L. and Kaufman, H.E. (1985) *Directory of Germplasm Collections* 1 II. *Food legumes (Soyabean),* International Board for Plant Genetic Resources, Rome.

Kajimoto, G., Shibahara, A. and Yamaghoji, S. (1982) Changes in the contents and compositions of lipids, fatty acids, tocopherols and sterols in soybean seed during maturation. *Journal of Japanese Society of Food and Nutrition,* **35,** 345–50.

Kokobun, M. (1991) Cultural practices and cropping systems for vegetable soybean in Japan, in *Vegetable Soybean: Research Needs for Production and Quality Improvement,* (ed. S. Shanmugasundaram), Proceedings of a Workshop held at Kenting, Taiwan, 29 April - 2 May 1991, Asian Vegetable Research and Development Center, pp. 53–60.

Lumpkin, T.A. and Konovsky, J. (1991) A critical analysis of vegetable and soybean production, demand and research in Japan, in *Vegetable Soybean: Research Needs for Production and Quality Improvement* (ed. S. Shanmugasundaram), Proceedings of a Workshop held at Kenting, Taiwan, 29 April - 2 May 1991, Asian Vegetable Research and Development Center, pp. 120–40.

Masuda, R. (1991) Quality requirement and improvement of vegetable soybean, in *Vegetable Soybean: Research Needs for Production and Quality Improvement,* (ed. S. Shanmugasundaram), Proceedings of a Workshop held at Kenting, Taiwan, 29 April - 2 May 1991, Asian Vegetable Research and Development Center, pp. 92–102.

Nakano, H. (1991) Vegetable soybean area, production, demand supply, and domestic and foreign trade in Japan, in *Vegetable Soybean: Research Needs for Production and Quality Improvement,* (ed. S. Shanmugasundaram), Proceedings of a Workshop held at Kenting, Taiwan, 29 April - 2 May 1991, Asian Vegetable Research and Development Center, pp. 8–16.

Nelson, R.L. and Deiyang Wang (1989) Variation and evaluation of seed shape in soybean. *Crop Sci.* **29,** 147–50.

Palmer, R.G. and Kilen, T.C. (1987) Qualitative genetics and cytogenetics, in *Soybeans: Improvement Production and Uses,* (ed. J.R. Wilcox), American Society of Agronomics, Madison, WI, pp. 135–209.

Pandey, R.K. (1987) *A Farmer's Primer on Growing Soybean on Riceland,* IRRI, Los Baños, Philippines.

Saravitz, D.M., Pharr, D.M. and Carter, T.E. Jr. (1987) Galactinol synthase activity and oluble sugars in developing seeds of four soybean genotypes. *Plant Physiology,* **83,** 185–9.

Shanmugasundaram, S., Tsou, S.C.S. and Cheng, S.H. (1989) Vegetable soybeans in the East, in *Proceedings of World Soybean Research Conference,* (ed. A.J. Pascale), Buenos Aires, Argentina, 5 March–9 March 1989, Association Argentina de la Soja, Buenos Aires, pp. 1979–86.

Shanmugasundaram, S., Cheng, S., Huang, M. and Yan, M. (1991) Varietal improvement of vegetable soybean in Taiwan, in *Vegetable Soybean: Research Needs for Production and Quality Improvement,* (ed. S. Shanmugasundaram), Proceedings of a Workshop held at Kenting, Taiwan, 29 April - 2 May 1991, Asian Vegetable Research and Development Center, pp. 30–42.

Shukla, V.K.S. (1990) Benefits, hazards in dieting lipids. *Inform.,* **1,** 150.

Shurtleff, W. and Aoyagi, A. (1989) *History of Soybeans and Soyfoods,* The Soyfoods Center, P.O. Box 324, Lafayette, CA, (in press).

Smith, K.J. and Huyser, W. (1987) World distribution and significance of soybean, in *Soybeans: Improvement, Production and Uses*, (ed. J.R. Wilcox), American Society of Agronomy, Madison, WI, pp. 3–22.

Takahashi, N. (1991) Vegetable soybean varietal improvement in Japan – past, present, and future, in *Vegetable Soybean: Research Needs for Production and Quality Improvement*, (ed. S. Shanmugasundaram), Proceedings of a Workshop held at Kenting, Taiwan, 29 April–2 May 1991, Asian Vegetable Research and Development Center, pp. 26–9.

Tsay, J.S., Lai, S.H. and Tsai, C.L. (1991) Present and potential cropping systems for vegetable soybean in Taiwan. In *Vegetable Soybean: Research Needs for Production and Quality Improvement*, (ed. S. Shanmugasundaram), Proceedings of a Workshop held at Kenting, Taiwan, 29 April–2 May 1991, Asian Vegetable Research and Development Center, pp. 65–72.

Tsay, L.M. and Sheu, S.C. (1991) Effects of cold storage and precooling on the quality of vegetable soybean, in *Vegetable Soybean: Research Needs for Production and Quality Improvement*, (ed. S. Shanmugasundaram), Proceedings of a Workshop held at Kenting, Taiwan, 29 April–2 May 1991, Asian Vegetable Research and Development Center, pp. 113–19.

Tsou, S.C.S. and Hong, T.L. (1991) Research on vegetable soybean quality in Taiwan, in *Vegetable Soybean: Research Needs for Production and Quality Improvement*, (ed. S. Shanmugasundaram), Proceedings of a Workshop held at Kenting, Taiwan, 29 April–2 May 1991, Asian Vegetable Research and Development Center, pp. 103–7.

USDA-FAS (1991) World oilseed situation and outlook. Circular Series FOP 5–91, United States Department of Agriculture.

Watanabe, D., and Kishi, A. (1984) *The Book of Soybeans*, Japan Publications Inc., Tokyo.

Wilcox, J.R. (ed.) (1987) *Soybeans: Improvement, Production and Uses*, 2nd ed., American Society of Agronomy, Madison, WI.

Wilcox, J.R. (1990) Varietal development for specialty markets – the niche varieties, in *Proceedings of American Seed Trade Association*, (ed. D. Wilkerson), Washington, DC, **20**, 6.

Yazdi-Samadi, B., Rinne, R.W. and Seif, R.D. (1977) Components of developing soybean seeds: Oil, protein, sugars, starch, organic acids, and amino acids. *Agronomy Journal*, **69**, 481–6.

Young, V.R. (1991) Soy protein in relation to human protein and amino acid nutrition. *Journal American Diet Association*, **91**, 828–35.

Index